RUNNING THE AMERICAS

AN EPIC SOLO & UNSUPPORTED RUN ACROSS TWO CONTINENTS

JAMIE RAMSAY

For privacy reasons, some names may have been changed.

First edition 2025

Book Cover by Aiden Barefoot
Edited by Jasmin Naim
Proofread by Natasha Wagner

Paperback ISBN: 978-1-0683739-0-9
Kindle ISBN: 978-1-0683739-1-6

Published by Jamie Ramsay

www.jamieramsay.co.uk

TABLES OF CONTENTS

PART ONE – NORTH AMERICA

PART TWO – CENTRAL AMERICA

PART THREE - SOUTH AMERICA

Imagery from the journey can be found at

www.jamieramsay.co.uk

PART ONE – NORTH AMERICA

CHAPTER 1 - GETTING TO THE START LINE

DISTANCE TO BUENOS AIRES: 17,011KM

A local reporter pulled me aside: 'So you're going to run 17,000km to Argentina solo and with no support whatsoever?' I nodded. There was an incredulous pause. Followed by, 'Why?'

It was a good question and one that I may not have fully thought through myself. In 2014, my life in London was spiralling in a direction that didn't sit well with me. It had become destructive rather than productive and this adventure was my way of altering that course. Was I overreacting and was it even achievable? It didn't matter, I knew I had to do something.

My London existence could easily have been mistaken for being near to perfect. I had been working for an International Public Relations business for nearly 12 years and had recently been promoted to Partner. My brother and I co-owned a lovely flat in Parsons Green, a nice part of London. Friends were plentiful and when I wasn't working, I was either to be found in the gym, out for a run, or socialising.

The truth behind the façade was very different. I felt trapped. Like so many people, I wasn't passionate about my job and it certainly wasn't something I wanted to be doing for the next three decades of my life. I had a sneaking suspicion that my recent promotion was due to an internal power struggle in the firm and less so my ability as a PR operator. The gym was where I went to hide and the pub was where I went to numb the boring predictability of my life. London had well and truly gotten its claws into me and provided a life that depended on pay rises and keeping up with those around me. I was living according to the pace of the herd and didn't feel at all comfortable with that. Yet the more I dreamt of breaking free, the steeper the spiral became and the less likely that freedom seemed.

I could well have become stuck in this self-perpetuating cycle of pity had it not been for a single night that made me realise that change was not an option but a necessity. It was a Wednesday night and I'd been out for far too many drinks with friends. When we stumbled out of a particularly seedy Piccadilly nightclub at three o'clock in the morning I should have boarded the night bus home, but instead I made the curious decision to head to my office and sleep there. On waking the next morning, I found myself on the cold, sterile bathroom floor in a crumpled suit, my jacket folded up into a pillow and with a splitting headache. I pulled myself up and took the proverbial long, hard look in the mirror. Let's just say I didn't like what I saw: bloodshot eyes, hair greasy and dishevelled and I could already feel drinker's guilt coursing through my veins. This jolt was my signal that life had to change. In that moment, it wasn't really clear what that would look like and I could never have foreseen where life would take me over the next few years.

Returning to the journalist's question, I found it hard to encapsulate my feelings in a few coherent sentences at the best of times. In all honesty, my mind wasn't fully focused on the interview either. Instead, I was mentally running through everything I thought I needed to do, very aware that this was brand new, so I was bound to have forgotten something essential, although had no idea what that might be.

In my efforts to warm up for what lay ahead, stretching was proving especially difficult with a camera pointed in my face. It wasn't just physical either; I was trying to prepare myself mentally and emotionally. Everything had seemed very different from the safety of a London pub. For possibly the first time since conceiving this idea eight months prior, it suddenly felt very real, very scary and a little overwhelming. It is very easy to plan for what you think will lie ahead but when you are about to take your first steps into an adventure that will take over a year to complete, there is a very different pressure, and strangely, it was one that I was immediately addicted to.

Getting to the start line had taken months of preparation, but in truth very little planning. After that fateful awakening on the office bathroom floor I bumped into a friend and colleague on my way to my desk – she tells me now that she noticed something different about me that day, a strange look in my eyes. I looked around at my older colleagues and realised for sure that I couldn't stagnate in this environment for another 25 years. Would I have turned out like the partner who drank too much, the one that never saw their family and/or those having affairs? I needed a completely new direction because passion was the missing ingredient from my current life. Nothing mattered more in that moment than making passion front and centre going forward. Having followed adventurers like Mark Beaumont, Ben Fogle and Sean Conway taking on the most awe-inspiring adventures for years, I'd always perceived a barrier between my existence and theirs. In moments of reflection, however, maybe the only difference between them and me was that they were braver and more willing to risk everything to live a more adventurous life, whereas I was simply observing theirs? With the flames of frustration taking hold, now seemed as good a time as any to straddle that gap and go in search of fulfilment.

One night over dinner, my best friend showed me a video of the late British philosopher Alan Watts in which he posed the question, 'How would you really enjoy spending your life?' At the time most of it had washed over me, but I'd soon come to realise that the words, 'What would you like to do if money was no object?' had affected me in a profound way.

Running had been an early passion. Childhood photos capture me running and I vividly recall feeling competitive at school sports days and Pony Club tetrathlons. But it had tailed off during university and my subsequent move to London, and it wasn't until a boozy afternoon with friends that I was roped into signing up for the

2004 Stockholm Marathon. I started the race thinking I'd stumble over the line in around four hours so was gobsmacked when I finished in three and a half hours. I remember crossing the finish line, finding a seat in the Stockholm Olympic Stadium, burying my head in my hands and letting a few tears fall down my cheeks as the huge sense of achievement washed over me. Even though I swore I'd never do another, I somehow found myself Paris-bound two years later.

On this occasion the excessively boozy night came the night before the marathon itself, when my sister, girlfriend and I consumed about seven pints of cider and a packet of cigarettes each. So it came as a shock to cross the finish line in a time of about three hours and 20 minutes. This would turn out to be my last proper marathon and having proved myself capable of finishing long distance road races, I started looking for the next challenge. Watching one of my heroes, Ben Fogle, run the Lewa Marathon in Kenya, I immediately wanted to emulate him. Although the Lewa Marathon is technically a full marathon, its trail route loops twice around a safari park and instead of dodging other runners or discarded energy drinks as I had in Stockholm and Paris, here there would be a whole new set of challenges. I found someone equally adventurous and signed up for the 2010 edition. My enthusiasm was unaffected when my partner dropped out last minute, so alone I boarded my flight to Kenya. I then drove my hired four-by-four from the Muthaiga Country Club in Nairobi to the Lewa Wildlife Conservancy where I seemed to be the only runner pitching a tent. While my fellow competitors enjoyed the luxury of safari tents, buffets and game drives, I sat on the floor and dined off instant noodles. After learning that the guys working at the event had to walk 10km each day to be there, I managed to fashion a makeshift game drive in return for chauffeuring a couple of them home. I may not have seen all the animals but I think I got a great taste of Kenya and more importantly, adventure. The race itself delivered everything I wanted and while my execution of the race may not have been optimal, I learnt so much and it fed my need for exhilaration.

This would be no isolated event, and three years later an appetite to push myself further re-ignited, so I enrolled for the first edition of the Jungle Marathon in Vietnam, a 240km six-day self-supported race. After months of training, preparation and fundraising and with just weeks to go, I received an email informing me that the race had been cancelled due to the organiser being ill. At first I was completely gutted that this had been taken away from me, especially as I was unable to afford the flights to China where would-be participants had been offered a free space in the Gobi March instead. After much reflection, I took things into my control and boarded my flight to Vietnam where I would set out to achieve what I had told people I would do, run 240km in Vietnam, unsupported. Vietnam was amazing on so many levels but the thing that stuck with me most was the feeling of happiness and fulfilment on reaching my self-imposed finish line. Undoubtedly, this feeling was the seed for what was to come.

How then to combine running with my other passion, travelling? I'd already travelled around the world twice and knew that the sense of freedom made me feel alive. While contemplating what running adventures may lie ahead, I turned to Ewan McGregor's *Long Way Round* and *Long Way Down* adventures and pondered what it would take to recreate them on foot, soon realising that I lacked the skills, finance or knowledge to achieve either route solo and unsupported. McGregor had travelled east from the UK to Alaska, and then south from the UK to Cape Town by motorbike. What if I tackled the Americas? Immediately, the route from Vancouver, Canada to Buenos Aires, Argentina shone out. It had everything I was dreaming of and feasibly, if I ran quickly, it could be done on a UK passport without any visa applications. Even better, it required only two languages. Later, people would ask why I didn't run from Alaska to Ushuaia at the tip of the continent, and the honest answer is that I simply didn't (still don't) care about firsts, records or getting my name into the history books. This challenge was for personal reasons and that sense of purity is important to me.

As you know, this was intended to be a solo and unsupported adventure. At the time, it wasn't clear what that would entail or that the non-running elements of the adventure would turn out to be the real challenge. What I did know was that, in addition to running roughly a marathon every day with all my kit, I would have to take care of all the logistics. These included finding a place to sleep each night, making sure I had enough food and water to fuel my body, as well as taking care of any mechanical and unexpected issues that would surely arise. Not to mention the challenges presented by humans, weather and terrain. One element I probably didn't fully appreciate was just how much the mental side of being solo was going to loom over me, especially when times got tough and injuries started to niggle. If any hint of doubt surfaced, nobody was going to tell me everything was ok. I was alone, and had to be prepared to be strong physically, mentally and emotionally.

Having decided to run solo and unsupported from Canada to Argentina, the next step was to plan the actual route. Now it might sound a little absurd, but this part took me all of ten minutes. I simply entered Vancouver to Panama City into Google maps and then Cartegena to Buenos Aires. A very rough route came up, and even though I knew it wouldn't be exact, the knowledge that a car would have been able to undertake the journey gave me some peace of mind.

Most people focus on the distance they'll travel and don't give much thought to obstacles along the way. My route would be the equivalent of running over 400 marathons, but I was to discover that the real interest would lie in some of the challenges I'd encounter. Early preoccupations included navigating my way around the Darién Gap, a 100-mile roadless stretch between Panama and Colombia, running through the Sechura Desert in Peru, crossing the Atacama Desert and the Andes Mountain range, especially the crossing from Chile into Argentina at 4,800m above sea level. This says nothing of the huge variety of weather conditions and the challenges they brought, especially the need to prevent dehydration or sunstroke and how to deal with freezing temperatures at night.

Financing the trip was my next dilemma. I knew I didn't have enough cash to finance the entire distance but estimated that my savings would fund the trip as far as Panama. It might seem naïve, but I firmly believed that if I set off with pure intentions, a positive attitude and armed with a smile then, if it was meant to be, the money would find its way. To this day, I still approach life in this way. I also didn't want to be that person who says, 'I would have done it but didn't have enough cash to finish so I never started.' Even if I had run out of finance, running from Vancouver to Panama would still have been an incredible achievement in itself that I could have been proud of. One that would most definitely be worthy of quitting my career for.

Packing took a lot less time than I thought it would, though looking back now I laugh at how much stuff I took. How much easier would it have been armed with the knowledge I have accumulated since? On the basis of a little research about long-distance running solo and self-supported, I stumbled upon Jamie MacDonald, a British superhero runner who had run 5,000 miles across Canada. He had started out with a backpack but ended up using a Thule running stroller, designed for pushing children, so I immediately contacted the Thule marketing team. To my surprise and delight, they responded, agreeing to provide me with a sleek lime green and black Cheetah two-baby stroller. With my stroller in place, I started to fill it with everything I thought I might need. The Scottish outdoor brand Vango kindly provided a tent with a porch for my stroller, two sleeping bags (one for North America and one for South America), a sleeping mat and a classic Trangia stove. Overboard supplied some waterproof bags and Powertraveller gave me some solar panels and powerpacks. As the kit rolled in, it dawned on me that I may actually pull off this potentially insane adventure. I had less response from running clothing and shoe brands so decided that I would just settle for what was already in my cupboard, knowing that I could just pick up new kit as I made my way through the United States of America.

Training for a run of this magnitude is almost impossible to get your head around. I knew I could run but to test my resolve I made myself cover 10km every day for a full month just to check I had the discipline to keep going, no excuses. On completing that, I decided to limit my training to more physical maintenance, believing that the running on the adventure itself would provide the bulk of the practical training. I did take the baby stroller for a few runs around London to get my body used to the different technique, and while undertaking some dangerous road crossings, was shouted at by passers-by who thought I was being reckless with a child onboard. There was nothing else for it but to attach a 'Baby Not Onboard' sticker to my stroller.

One of the interesting dilemmas I had to come to terms with was how I viewed the impending adventure. To get brands onboard, drum up interest and raise money for charity, I had to focus on all the hardest elements, such as the mileage, loneliness and obstacles like running across deserts and over mountains. I had to make it sound as hard as possible. But at the same time, I had to play the whole thing down in my own mind so I really believed it was all achievable. This tricky balance of perception was something I would have to manage throughout my adventure; keeping the allure of the adventure for others while normalising it for myself.

Raising money for charity was an important element of my adventure and I had initially chosen two charities to support. The first was Macmillan Cancer Care and the second WaterAid. Macmillan had helped care for a relative and I knew that the odds were that cancer would most likely rear its ugly head again. The next choice was WaterAid because I knew that a constant supply of clean water was going to be one of the most precious resources of my adventure and it should be a basic right for all humans. CALM, or the Campaign Against Living Miserably, was a last-minute addition to my charity list. My good friend and then colleague Jo invited me along to her fundraiser for the charity. I'd never heard of CALM and had very little awareness of male suicide at the time, but when I heard the speakers talking about how many young men feel that the only way to deal with their demons is suicide, I knew I had to support this amazing charity. It didn't escape me that I too had been in a downward spiral, but with support and the necessary resources had found an alternative path. The message that there is always another way and removing the stigma of suicide remains a cause worth supporting.

While all this planning and preparation was going on in the background, I had one vital thing to take care of… quitting my job! In my head this would have been done in a suitably dramatic way and I would have been marched out of the office, my belongings bundled into a cardboard box, but the reality was far less heroic. The decision to make this drastic change was made in November, but it wasn't until the following April that I would inform my boss. Even then it was he who had approached me rather than the other way around. While spending hours daydreaming and planning my big adventure, I wasn't really concentrating on my day job. One day my boss approached me and asked if I would step into one of the small meeting rooms that overlooked the office. Out of genuine concern, he asked if everything was ok, as he had noticed I had been distracted. Putting my fear aside, I just blurted, 'I need to hand in my notice.'

'Are you leaving to join another agency?' he asked with surprise.

'Not exactly, I'm going to run from Vancouver to Buenos Aires.' His surprise turned to incredulity and he quickly pulled the other managing partner on our team into our discussion.

My fears about how they would respond had been for nothing, as both were amazing in their response and went above and beyond to ensure that I was looked after in my departure. Their only request was that I would work until the end of July, to which I agreed, and not to tell any client until an appropriate time, which I was less good at. In fact, in my excitement I had already shared my news with one of our largest clients, and for weeks he had been checking in before meetings to see if I had told my bosses.

The sense of relief that my truth was out was immense, and it meant I was free to set a date for my adventure—the 15th of August 2014. There was no significance to this date other than it was two weeks after my last day at work. I worked out my notice, started putting my affairs into order, then had an appropriate amount of leaving drinks.

My last night in the UK was spent eating, drinking and smoking to excess with my brother, his wife, my sister and her boyfriend. Incredibly, before going to bed that night it struck me that I hadn't even thought about insurance. Not knowing what cover I would need, I opted for the standard bronze backpacking package, honestly believing that in essence that was all I was setting out to do.

The next morning, standing outside Gatwick Airport dressed in my running kit and pushing a baby stroller, I lit my last ever cigarette before boarding my flight to Vancouver.

CHAPTER 2 - 17,011KM TO BUENOS AIRES

DISTANCE TO BUENOS AIRES: 17,011KM

As I stood in the shadow of Vancouver City Hall there was to be no bell, whistle, or official start but when the time felt right, and people seemed anxious to get to their jobs, I took the first steps of my life-changing solo run to a city on a completely different continent.

A ripple of cheering broke out from the small crowd mostly made up of people I didn't know, that had formed to bid me farewell. I was feeling excited but found the whole experience a touch surreal. These were the first steps of a huge expedition that, with luck, would end thousands of kilometres south and be the catalyst to change my life completely. The planned 17,000km distance had an unreal quality but I didn't want to freak myself out at the very start of the trip. Deciding not to think about the total number, instead I focused on each individual step. 'Take it one day at a time,' I told myself. 'All you have to concentrate on is getting to the end of this day.'

A friend of a friend had offered to run the first few kilometres with me. It was our first meeting, however, so it felt a little strange sharing something so momentous and personal. I was glad of his conversation though, as it helped to keep my mind off the fact I was pushing my fully loaded baby stroller for the first time. We discussed the history of Vancouver, how he knew my friend and all the competitive runs he was training for. All I could do was run and try to keep up. At the outskirts of Vancouver there was no turning back, and we were now running south towards the border with the USA. Many people I've told don't believe this, but I truly had no idea where I'd sleep that first night. My policy of not organising places to stay each night was deliberate, in the belief it would give me the flexibility to change my plans as I went along if necessary. This was nonetheless somewhat nerve-wracking, but I felt strongly about not setting myself concrete targets to hit every day. That bit of flexibility allowed me to perform to whatever level was achievable on any given day, but was also an excuse not to force myself to continue pushing when I didn't have to. For this adventure my philosophy would be minimal planning, as anything more would detract from the very essence of what an adventure should be. This was an alien feeling because generally I like to plan and have everything in order. Now, however, I was giving myself permission to be looser with the logistics.

It was late afternoon when I trotted into South Surrey, grand houses nestling between parks in this palpably lush and affluent suburb. Tonight was assigned to wild camping but to be honest, the manicured nature of the area alarmed me. In my mind's eye there'd be a rough and ready campground or common land, but this seemed more like pitching my tent on somebody's lawn. My innate Britishness initially stopped me from just pitching camp and making myself at home, but, well, that didn't last long. If I wasn't prepared to be brazen, the next couple of years would be really hard. Strolling around for what seemed like an age, it dawned on me that I was possibly less carefree than I'd imagined. Not wanting to be moved on meant having to discount a lot of possible sites. It would have been so embarrassing to have made a foolish mistake on my first night!

After an hour or so, I found a small, isolated corner of the park and set about making camp. There weren't too many houses around, but I was still worried about being moved on by the local police. I had no idea what they would make of my journey or how they would react. If I had been in the UK there is no chance I would have been so bold, but somehow being an adventurer on a mission seemed to give me permission to push the boundaries of what I thought was proper.

While setting up my tent, I realised that I'd never actually checked that my running stroller would fit inside! Luckily, and by some miracle, it fitted perfectly in the front porch section after I had removed the quick-release front wheel and the handlebars. Turning my attention to cooking, I set up my Trangia stove, my mouth watering at the prospect of pasta and pesto. For safety reasons I hadn't been able to bring methylated spirits with me on the plane from the UK and, amid all the excitement of setting off, had forgotten to buy any in Vancouver. Instead, I bought some camping fuel from a store on the side of the road earlier that day, and when I went to light the stove, it dawned on me that I may have bought the wrong fuel as I couldn't detect a flame. Even after adding more fuel still, no heat radiated. I casually sprayed some more fuel and all of a sudden, flames burst from the stove and started creeping across the grass towards my tent. In a vain attempt to extinguish the fire, I started bashing the spreading blaze with my flip-flops, setting them alight in the process. Panicking, I tried to douse the fire with water from my supplies but that only made things worse. Thankfully, it finally started to subside. As I sat back and tried to get my heart rate back under control, I heard a cough from the other side of my tent.

An elderly couple greeted me as I stood to face them. It turned out that they lived nearby and had popped over to assess who had set up camp. 'Are you alright?' the woman asked in her soft Canadian accent. I was incredibly flustered but tried to explain as coherently as possible what I was doing. 'I'm running from Canada to Argentina,' I said. 'I've just begun,' I added, possibly unnecessarily. The couple nodded understandingly.

'What do you think about hellfire?' said the husband. This was unexpected.

'I'm sorry, what?'

It turned out that my temporary neighbours were Jehovah's Witnesses who were keen to spread the word wherever the opportunity presented itself. Luckily, they seemed unaware, or were at least unbothered, that I had just dealt with another sort of fire in their park. Tired, disoriented and a little frazzled, this didn't seem the ideal time for an in-depth theological debate but thankfully they seemed keener to talk than listen, so I just sat back while they proselytised. When they were done, they gave me a church leaflet depicting a family farm setting with a couple of children cuddling a panda and generously asked if there was anything I needed. Amazed, I jumped at the offer and requested water, having just wasted my supplies taming the fire and not wanting to risk leaving everything to find more. They kindly obliged and soon returned with bottles of water, a small bag of supplies and a $20 bill slipped in for good measure.

As I sat in my tent chewing al dente pasta, the day's events filled my mind. Even if the start to my adventure hadn't been as smooth as it could have been, the generosity and warmth of people was clearly going to play such a vital role in what lay ahead. I was tired but exhilarated. My adventure had finally started and after surviving my first day, nothing was stopping me from continuing south. I slept like a log and the following morning it seemed advisable to buy breakfast from a café rather than risk another potential wildfire.

Although this was only my second day, my first border crossing, from Canada to the United States, was imminent. I took the most direct road, running along the hard shoulder of Highway 99 just a few metres from the traffic. From the outset I had decided to run with the traffic rather than against it—my rationale being that if I was going to be hit by a car, I'd rather not know about it—and it seemed a lot less stressful having cars overtake me rather than coming towards me at high speed. It was soon obvious the border was close, as cars were backing up on the highway with bemused looking people staring from car windows as I jogged by. Occasionally I got a whoop of encouragement.

I was a little sceptical about how the border police would react to me arriving on foot pushing a baby stroller. At the immigration office I was confronted by a tightly-packed line zigzagging across a large hall. Most were day-trippers visiting Canada. When an official told me I had to leave my stroller in the corner, I rather abruptly told him that wasn't going to happen because it, and everything inside it, was of vital importance to me.

That may sound dramatic but without these things, my adventure could have been over before it had barely begun. Thankfully, one of the counter staff overheard the conversation and ushered me over, asked for more details of my run and agreed to process my visa application. It transpired that I had inadvertently bought the wrong visa, but the border patrol officer worked with me, and I was granted the correct version on the spot. I did get a few quizzical looks as I reassured him that I would be able to run the 2,900km to the Mexican border within the 90-day permitted period. Pointing to my stroller, the official asked if I had any illegal items. Having no idea what would be classified as legal or not I was anxious to get moving. 'No,' I said confidently, and at that, he stamped my passport. With very little interference or checking of baggage, I was unleashed into the second country of my adventure feeling like I had taken a great leap forward and was now on a mission to run the entire length of the United States within 90 days.

CHAPTER 3 - DRUGS, A KISS AND POLICE

DISTANCE TO BUENOS AIRES:16,960KM

Running into the United States was a big deal for me. It may only have been 55km into the adventure, but it marked a huge milestone. It's curious how invisible lines on the ground make really noticeable differences. The Canadians had been very kind and supportive, if somewhat reserved. I had been amazed by the generosity of some waitresses who had clubbed together to collect money for my charities while I ate lunch on my first day. However, arriving in the US turned the dial right up. On my first visit to a supermarket the check-out lady asked what I was doing. When I explained she announced it to the whole shop over the loudspeaker system only to receive cheers and shouts of support. Straightaway I knew the USA was going to be a far more interactive experience.

My first stay in the US was to be in Bellingham where Kathi, the mother of a friend, Alina, had thoughtfully offered a bed for the night. Although only two days into the adventure, the idea of seeing a friendly face and having a nice warm bed for the night was very appealing, even though it was a few kilometres out of my way. Not wanting to tire myself out for the second day in a row, I broke the day up into manageable chunks interspersed with regular coffee refills and cakes.

Of course, Kathi's house was situated at the top of a long hill, but when I arrived she was waiting with open arms. She knew how to greet a tired runner, showing me straight to my bedroom, which turned out to be a whole apartment attached to her house.

After a quick shower, I felt more presentable, which was just as well as we were heading straight out to a dinner party. My mind was a little frazzled so keeping up with the highly intellectual dinner guests was tricky. A mixture of anthropologists, historians and archaeologists shared an astonishing array of knowledge about the countries I was going to be running through—and their kindness and enthusiasm about what I was attempting helped me get through the night. I sat back, enjoyed the food and let the conversation wash over me. Even so, I couldn't fail to notice one recurring theme—and it would surface many times over the coming weeks—namely how wary they were of Mexico. Everyone around the table warned me of the dangers of travelling there and advised against it. I knew that to run from the top of one continent to the bottom of another you could hardly skip a country, especially as in my mind, Mexico was one of the least problematic! Looking back, I am glad I didn't overthink what each country had in store for me. I had made a calculated decision not to research the route that lay ahead as I wanted every day to be into the unknown. It is the unknown that makes the adventure and everything I could do to preserve that just helped to make the journey more thrilling. I maintain this approach in every adventure and some could say in life.

Kathi's generosity continued with a hearty breakfast and care package for the road, including blister plasters and pepper spray just in case I came into contact with bears, something I had not even considered was a possibility! The stand-out gift was instant coffee sachets. It was a small thing but it really would revolutionise my morning routine going forward.

For the third day in a row, my inexperience was evident in the face of a relentlessly long uphill section. Rather than sensibly walking my stroller up the steep sections, I tried to run. Running every step of the way felt necessary because so many seemed to doubt I'd make it more than a week into the trip. Even though nobody was physically watching me, I felt under the microscope and walking—even a short stretch—was somehow unacceptable. This puritanical work ethic quickly revealed itself as an error. Gradually through the day, something was clearly not quite right and as I descended the other side, a pain in my Achilles slowly started to develop. My running transitioned to a jog, then a walk and finally a hobble.

The cause of the injury was clearly a dramatic increase in mileage, the exact reason I had originally planned to ease into this expedition. What with running up hills, charging off at the start far too fast and unwittingly running more each day than intended, this was the exact opposite of what I'd planned. I was well aware that Achilles injuries could be longstanding and difficult to cure, especially if still moving, and I immediately cursed myself for potentially putting the whole expedition in jeopardy at such an early stage. Had I just quit my career, rented out my flat and flown to the other side of the world to scupper everything due to inexperience?

Determined not to let it get the better of me I continued at an easier pace, stretching my Achilles whenever possible. The surroundings were spectacular, so focusing on that helped take my mind off the negativity building inside. Positivity is my natural state, especially if I just have myself to rely on, so this glitch couldn't be allowed to derail me. At that moment, a large convoy of motorcycles came roaring past, each bike carrying a young child on the back, each with differing disabilities and each smiling from ear to ear as the wind rushed across their faces. I spoke to one of the bikers who had pulled over to the side of the road and he explained that they were a charity who tried to help children in hard circumstances have a distraction for a while. 'Stop complaining,' I told myself. 'Just keep moving forward.'

I took the next few days a bit easier. The adventure was still all so new, and everything filled me with wonder. All these challenges were firsts, and unforeseen encounters reinforced the sense of adventure. For example, I came up against a 'road closed' sign and assumed that this must only apply to cars. After a couple of kilometres, it was clear that the road was indeed completely blocked as a three-lane bridge was being rebuilt. This was a construction site and no place for a runner. The first workers that saw me turned me away without even a glance. On enquiring, I was informed there was definitely no way through. Dejected and feeling defeated, I turned around to retrace my footsteps and take the longer diversion. Before I'd gone far, a young chap called me over. He had heard what I said and was convinced we could navigate over the bridge if we worked together. We cautiously scaled the basic scaffolding and were soon tiptoeing along the edge of the bridge, squeezing past busy welders. My heart was in my mouth as we gripped my stroller over the barrier with nothing between it and the raging river far below. After arriving safely on the other side, we took a couple of selfies and he went back to work. These acts of kindness were going to be the fuel I needed to get through this running adventure—determination alone was not enough.

At North Seattle my host was going to be Alina's sister Larissa and her family. Reaching there slightly early, to avoid disrupting their life any more than I was already going to, I found a bar and had a couple of drinks to avoid being a burden on people I'd never met yet were willing to help me. On arrival I was given the warmest welcome, very similar to that of her mother a few days earlier. The youngest daughter, Isla, had drawn a big chalk picture on the wall of my bedroom that read: "Welcome to Seattle JAMIE." It was the cutest thing. We had a nice home-cooked meal and shared a lot of fun stories. The next day I was given the largest bag of pumpkin muffins, that I think were meant to last me a few days but vanished very quickly.

It seemed a shame to just circumvent Seattle, so I ran into the centre to get the feel of the place and visit some of the more touristy destinations such as the Space Needle and the very first Starbucks. In the famous fish market, I had the best fish sandwich, but more importantly, met an American couple who had spotted the cancer charity logo on my tee-shirt. He was a large, friendly guy with a massive white beard, while she was much thinner with short hair and, as it turned out, had cancer. They were in Seattle to seek treatment and she accessorised a big smile with a tee-shirt that read: 'Kicking Cancer's Butt—what's your superpower?'

We enjoyed a great conversation, and they were so inspiring with their attitude that positivity is vital when fighting such a huge battle. I was glad that I'd chosen to wear the charity tee as it was a way in for people to talk to me and I was getting so much from hearing about other people's incredibly different challenges. Life as a runner was opening me up to all sorts of interesting conversations that my previous self would have been too busy or egocentric to have.

It was time to hit the road again, and the best way to get from Seattle to the Pacific Coast was taking the ferry to Bremerton. Not only did it avoid the busy freeways, but it allowed me to enjoy the beautiful sights of Puget Sound.

Strolling out onto the deck of the Issaquah ferry, I got talking to a guy of a similar age, with tattoos covering every available part of his skin and muscles that stretched his tee-shirt. On first impressions, he looked scary and someone whom I might normally have avoided but as he was holding the door open for me, conversation sparked. He introduced himself and after enquiring where I was from told me he was heading home to Bremerton. As an ex-gang member he'd just served 10 years in a state penitentiary. Bremerton was, he told me, a town ravaged by drugs, and methamphetamine in particular. Allegedly, drug gangs roamed the streets and there were walled communities where unschooled children ran around freely. This was a whole new world for me. I'd barely even heard of meth, so this was my first exposure to its effects on people and communities. The stories I was hearing were so alien and in such stark contrast to the beautiful scenery.

The man was very open about his experiences and showed true remorse for his actions, including a life of crime and several racist tattoos on his body. His plan was to return to Bremerton to try to inform younger people of the consequences of a life of drugs. As the Seattle skyline retreated into the distance he said, 'I'm going back home to tell young people how their life will unfold if they follow this path—either in jail or dead.' He was full of zeal and seemed genuinely inspired to save others.

As we disembarked, an older lady saw my stroller and asked what was I doing and for which charities was I raising money? On hearing that male suicide was the main focus, she told me that her son had been in a relationship with a young man and, while she found the relationship difficult, she'd accepted his choice. Tragically, her son had taken his own life a few years before, which she attributed to being unable to communicate how he felt. The boyfriend had then taken his own life too. She was distraught that neither of them felt able to reach out for help.

After hearing such harrowing stories, I now found myself running through some starkly beautiful countryside and awesome woodland. Possibly due to hearing about other people's traumas, I now seemed to appreciate every moment more, no matter how small.

Finding somewhere to sleep that night was tricky. This was only the start of week two and I was still getting used to proper rough camping. After about 50km a park and ride came into view as the sun was setting. On the far side next to the woods was a flat space of grass that looked perfect for my tent.

Within a few minutes a large pick-up truck pulled up and the driver waved me over. A friendly-looking chap hung out the window and asked what I was doing. I gave him a quick overview and asked if he thought it was okay if I camped there. He winced and said that it may not be ideal as while it looked tranquil in the daylight, at night this spot became a different world. His first concern was a homeless man who lived in the woods. He described an aggressive drunk whom everyone avoided and who occasionally emerged from the woods. His second worry was that this was the local hotspot for meth dealing. We discussed the pros and cons and decided that while the place wasn't ideal, if I kept myself to myself then there probably wouldn't be too much bother.

We then discussed my chosen charities and CALM in particular—it turned out he had his own story to tell. Having been born and raised in a poor part of New York, he'd been involved in gang activities—not by choice—as in his community you were either in the gang or against it, and if you were against it, you were the enemy of it. Thanks to this he ended up spending a year in jail, which would haunt him for the rest of his life. Although he managed to get out, find a job and get married, he later lost his job and found life increasingly difficult to come to terms with. He'd been raised to understand that being a man meant being the strong figure in the family, and if he was not providing for them then he was not fulfilling his role as a husband. This led to depression and even thoughts of suicide.

Thankfully, during his dark times his wife stuck by him and made sure they focused on the important things. They had moved out to Shelton to live a quieter life. Tailing off, he wished me luck in my endeavours and drove home, only to return later with a little care package of water, toilet roll and vegetables from his garden. While he thought it wasn't much, to me it was an amazing token of generosity.

While brushing my teeth there was a rustle from the bushes. Clearly, it was a person. I made the split-second decision to investigate and confront the man from the woods, rather than hide away. I waited as the figure slowly emerged from the bushes. He was about 50, unkempt and sported a bushy beard, but under all the dirt lay a beaming smile. We made our introductions and began to exchange stories. I spoke of my run and he informed me that he had spent the last 30 days in a rehabilitation unit (a stint that had clearly not worked). He explained that he knew many people viewed him as just another homeless person, however, he preferred to think of himself as someone who had decided to sleep in the woods away from society. His daily routine involved riding the bus to the local town during the day and occasionally writing for the local newspaper. He'd had a hard life and told of his late wife and son, who had both taken their own lives. Spiralling towards depression and addiction, he too attempted suicide while in Mexico. When things had got too much to bear, he had consumed a whole bottle of brandy, swallowed a healthy portion of dog tranquillisers, then walked straight into a desert. Fortunately, his attempt was thwarted when he was found two days later and revived. He talked of his regret at ever thinking that suicide was the best solution to his problem. Ever since, he told me he'd focused on living instead. After about 40 minutes he crashed back into the forest to his camp and I returned to the warmth of my tent.

At about midnight, I was woken by headlights shining into my tent and the sound of an idling engine. The first car was soon joined by a second. There was no question what was going on thanks to my earlier warning, and while car doors opened and closed, I remained as still and quiet as possible. Hushed voices conducted their transaction as I sat in silence. The second car left and was soon followed by the first as a sigh of relief washed over me. It turned out to be just the first of a few transactions, but my presence was evidently not an issue so after a while I merely rolled over when another car arrived. While they couldn't have missed my tent when they drove up—the reflective bits would have lit up like a Christmas tree—thankfully both they and I decided to leave each other alone.

Over the next week, I had lots of experiences that helped me settle into this new life on the road, including a rather awkward moment in a bar in a village called Artic. I had stopped at a quirky little family-run campsite that was conveniently situated near a bar. After 10 days on the road and running over 350km, now felt like a good time to rest. My initial plan had been five days on and two days off, but that had gone out the window. I popped into the local bar, had a few drinks and tried to mingle. While chatting with a couple of ladies, one of them offered to take me out to breakfast. Naively, I accepted and the next morning she duly arrived at my tent and drove me to the town of Aberdeen, which incidentally is where Kurt Cobain was born. We had a delicious meal and a lengthy chat, which was a little surreal as I had only just gotten used to being very alone. That evening, I revisited the bar only to find my breakfast companion there once again. After a few drinks she insisted on walking me back to my tent and there was some nervous hanging around before a rather awkward kiss. Luckily, my tent situation prevented anything going further.

Heading south to Astoria, my Achilles flared up and running was proving painful. My day became a little brighter after meeting a chap who was cycling with his cat. His bizarre predicament made me feel rather normal. He had quit his job and cashed in his pension and was just cycling around in search of something but wasn't entirely sure what that might be. It made me thankful to have purpose and a destination. Before we parted company, he donated a portable speaker for my stroller as he was concerned that my listening to headphones was just too dangerous. This revolutionised everything.

Astoria was the next big town and getting there turned out to be my longest day so far. After leaving a small campsite that morning with pain shooting up my Achilles, I wanted nothing more than to prove to myself that I was able to control the pain and keep up my momentum. The scenery and weather were perfect, and with Brandon Flowers blasting on my new speaker I tried my hardest to distract myself from the negatives. Like most days, I didn't know where I was heading and just pushed on. As the day wore on my competitive side fired up, such that the prospect of covering 65km and making it to Astoria became a challenge that was impossible to ignore. My mind shifted from a feeling of despair in the morning to huge determination by the afternoon, but I was learning not to get caught up in a single feeling. The only way was to continually move forward and adjust accordingly. As I dropped down onto the banks of the Columbia River, I was treated to a spectacular sunset and a challenge I had never expected.

The Astoria-Megler bridge is the longest truss bridge in North America and the huge metal structure spans 6.5km over the Columbia River. I arrived just as the sun dipped into the Pacific Ocean. Having run nearly 60km, I was keen to rest, so my first thought was to leave the bridge crossing until the next morning and sensibly attempt it in daylight. On seeking somewhere to camp, I found the options very limited. At a small permanent campsite where people lived rather than visited, the welcome was frosty and soon had me returning to the bridge. My only available option was to cross to Oregon to the sparkling lights of Astoria. Although the last rays of light had disappeared, I could read the huge sign that clearly stated that pedestrians were not allowed to cross. I hadn't anticipated the possibility of not crossing nor considered an alternative. My maps showed a 120km detour to get around to the other side, which to me was simply not an option. This made up my mind to attempt the crossing under the cover of darkness and hope for the best. Positioning lights on my stroller and cap to give the appearance of a bike, I cautiously made my way onto the bridge. There was something hugely invigorating about being on such an enormous bridge at night and that childlike feeling of breaking the law. I made as quick a pace as possible, though my body was screaming as I ticked over the 60km mark for the first time. The lights on the other side enticed me and mentally I started making plans for a rest day. As I climbed the steep hump on the Oregon side, blue lights flashed from behind and a huge black truck pulled up, indicating that I should pull over. 'You are not allowed to be here,' a police officer shouted as he strode towards me. Feeling a little nervous, I knew I had to stand my ground and appeal to his better judgement.

'I just need to get to the other side of the bridge and if you make me get in your truck then my whole 17,000km run will be void, everything I have worked towards will be for nothing,' I stated, laying it on quite thick. You could see him struggle with what to do as he surveyed my running stroller and the laminated poster attached to the front.

'Ok, I'll tell you what we'll do,' he began, 'you continue to run and I will track you from behind and when we get off the bridge, pull over so I can take your details.'

Relief washed over me as I set off with the flashing lights of his truck right behind. More than relief, this was a feeling of surreal jubilation that my day was ending in this bizarre and exhilarating way. Pulling over to give him my details, I felt a tingle of illicit pride before going in search of somewhere to sleep.

CHAPTER 4 - WARMSHOWERS AND STRANGERS

DISTANCE TO BUENOS AIRES: 16,500KM

After resting in Astoria and visiting as many of the craft beer breweries as possible, I was ready to continue south. In Seaside, I got my first taste of Warmshowers, a hospitality exchange platform that connects touring cyclists with hosts who offer free accommodation. Allan, a cyclist I had met a few days earlier, had contacted someone who lived there to explain what I was doing. Despite being out of town, he was more than happy for me to stay over. He told me where his keys were, to let myself in and help myself to anything. Blown away by this generosity, I found a beautiful house with a big smiley face spray-painted onto the garage door. Being in a proper house was such a good feeling and I made the most of home luxuries. The only price for staying was to write an answer to this question in his guest book: 'Are you realising your dreams?' I wrote:

'Every day surprises me. You are an example of a truly kind person who shelters people from the elements. I am on day 16 of a 600-day expedition which will see me attempt to run (walk when necessary) from Vancouver to Buenos Aires. I am raising money and awareness for several charities: Macmillan, which supports cancer patients, WaterAid, which focuses on clean water provision and CALM which seeks to prevent male suicide. Each day I meet new people and there are always links to make and stories to hear. And realising my dream? The simple answer is that I am realising many dreams but also creating new ones. Each day is different and rewarding in its own way. Staying in your home is a fantastic example. Humans were made to travel and interact. That's how we learn to be better people. Thanks again, Jamie.'

The next couple of days fell over Labor Day holiday weekend and the traffic increased as I pressed on through Oregon via Garibaldi and Tillamook. After the famous cheese factory, I left the 101 Highway to take advantage of the quieter and more scenic coastal route along the shores of Netarts Bay Shellfish Reserve to Cape Lookout State Park, where I spent a night camping in the sandy woods with many others who had flocked to the beach for the weekend.

Neskowin was a pleasant surprise after a couple of days' running in the hot sun. A cold beer in Hawk Creek's Café on the Creek was irresistible after my thirst was awakened earlier that day at the Pelican Brewery. Ordering a drink at the bar, I sparked up a conversation with the man sitting next to me, who it transpired was the owner. Conversation became friendly banter, and he declared that I could eat and drink whatever I wanted for the rest of the day, on the house. After I tentatively ordered another drink, my host immediately stepped in to encourage me to sample all the craft beers, taste the wines and have sticky toffee pudding after a large Hawaiian pizza. He introduced me to his customers and their genuine interest in my adventure filled me with warmth and renewed resolve.

A couple got up to leave but turned back to ask where I was heading the next day. They lived in Depoe Bay, only 40km away, and we agreed I should swing by for the following night.

However, no offer of a bed for that evening was forthcoming, so with a full belly and slightly drunk, the beach below seemed a good place to camp. This turned out not such a clever idea as the sand proved astonishingly uncomfortable and a heavy dew made it stick to everything. Finishing the leftover pizza for breakfast, the congealed sand made a Sisyphean task of getting my stroller back to the highway. The only thing keeping me positive at that moment was the prospect of a nice clean bed that evening. This email from my imminent host reinforced that:

‘We enjoyed meeting you at Hawk Creek this evening and are looking forward to seeing you tonight. Ken is planning to cook salmon for you for dinner, which he just caught yesterday in the ocean.’

Arriving early in the charming coastal town of Depoe Bay, which claims to be ‘the world’s smallest harbour’, it seemed prudent to kill some time before burdening my hosts. A crowd of tourists surrounded the seawall that overlooked the picturesque harbour, so I went to investigate. In the water below a whale gracefully idled, mesmerically surfacing and diving for the next hour or so.

After a quick beer in the local pub, I jogged to Joann and Ken’s house and received an amazingly warm welcome. My room came complete with an ensuite bathroom, clean towels and fresh soap. Home comforts enveloped me and the red wine and salmon promised for dinner were particularly enticing. Joann and Ken had invited their neighbours over to meet me and to provide some security, and we enjoyed lively debates about the politics of their local community, guns and general American life. Being surrounded by normal people just talking about normal things was comforting, and a welcome break from the more challenging road. The next morning we regrouped, and after a long beach walk together they sent me off with a delicious packed lunch. These fleeting connections were so special to me, and I have tried to keep in touch with Joann and Ken and others like them.

CHAPTER 5 - HASH BROWNIES AND THE GOLDEN GATE

DISTANCE TO BUENOS AIRES: 16,300KM

The focus for the next seven days was keeping my head down to get as many miles under my belt as possible. Thankfully the campsites along the route were improving, as they were geared for cyclists on Route 101. This famous Pacific Coast Highway provides stunning coastal scenery that features rugged cliffs, sandy beaches, lush forests, and picturesque ocean vistas. The campsites were perfectly distanced and formed clear targets for each day and made getting into a routine a little easier. Routine was becoming the backbone of this adventure and the more I stuck to it, the easier the running got. I also hit a big milestone when entering California. At this point in the adventure, crossing state borders was still thrilling, besides marking milestones that maintained my motivation. Arriving in California filled me with a little extra excitement as I pictured the great cities of San Francisco, Los Angeles and San Diego.

Unbeknownst to me, a mini adventure was about to unfold in Crescent City. Allan (who had sorted out my accommodation in Seaside) and I, were about to spend some proper time together rather than just chance run-ins on the highways or at campsites. Allan was a slim man in his sixties who was on a cycling adventure. Whenever I think of him I remember the huge smile on his face and his love for the adventures his cycle touring provided. His trip was motivated by just having been given the all-clear from throat cancer and wanting to celebrate this through adventure. We immediately recognised a common bond and hit it off. Meeting as planned in Crescent City, we had found a Warmshowers host in a small church community centre. One of the main volunteers told us that the centre had only recently started accepting cyclists after a young chap had turned up on their doorstep. At first, they'd been sceptical about letting a stranger in but after inviting him for dinner, the stories he told helped to convince the rest of the committee to make this a more accessible service. I for one, and certainly many other travellers, will be forever grateful they made that decision.

The next day Allan told me about a series of fun runs in the Prairie Creek Redwoods State Park a couple of days later. Intrigued, I made my way to the visitor centre on the outskirts of town to find out more. While it might seem a little crazy and verging on masochistic to think about running even more than I was already, I'm a runner at heart so the race had an inescapable pull on me. I tracked down the organiser and explained my presence there. While interested in the idea of me joining the competitors, she wasn't convinced that I'd be able to cover the 55km to get there in time for the start of the race. We struck a deal. If I could get there in time, I could enter for free. Never one to turn down a freebie or back away from a challenge, I now had something new to focus on.

Between me and the race stood the longest hill of the expedition so far. As was becoming my default mantra, I reminded myself that there was a lot worse to come in South America and if I couldn't conquer this now then what chance would I have then? This hill was a mere 350m ascent; when I got to South America, I would have to overcome mountains well over 4,000m. Proving to myself that I could manage this would re-affirm my capacity to tackle bigger obstacles further down the line. The aim was not to find it easy but to manage how it affected me, both physically and mentally.

Getting to the Redwood Forest proved more difficult than I had anticipated and by late afternoon I started to suffer. As if by magic, Allan and his unrelenting enthusiasm appeared behind me, like a grizzled guardian angel. He slowed down and supported me until we arrived at the State Park, as the evening sun slid behind the majestic trees. It was game-on for the fun run.

Rising even earlier than usual, I was surprisingly full of pre-race jitters despite being excited to be running with other people for a change! On presenting myself at the start line, the lady remained true to her word and handed me my running number. Part of me wanted to enter the half or full marathon but I didn't want to risk more injuries, so I opted for the 5km. However, to make it more challenging, I decided to run with my running stroller. While being interviewed by the local newspaper, Allan jogged over. We hadn't talked about him taking part, but he just got an urge to run. He had never done a proper 5km run like this so thought doing it with me would be a fun introduction. I felt proud to run with him, especially after everything he had been through.

Because of Allan's cancer treatment, he had to spit continually. However, the path was tight, requiring everyone to run single file, so each time he spat, it would inevitably hit the unfortunate woman behind! She understandably got a little irate. When she aggressively confronted him, Allan calmly replied that it was due to his recent cancer, which thankfully appeased her. Barriers so quickly spring up when we don't fully comprehend someone else's situation, but equally they can disappear.

That evening a fun assortment of cyclists, hikers, hitchhikers and me gathered to cook together and eat around a campfire, drinking cold beers the hitchhikers had bought from the local store. A late arrival was a cyclist named Brooks, an American making his way north up the coast. Unbeknownst to either of us at the time, we would share a few more adventures in the months to come.

Leaving the jollity of the campsite, the struggles of relentless running started to tug again at my consciousness, partly due to the huge mileage but also the injuries and mild boredom that accompanied it. It's not that I was losing interest in the adventure or enjoying it less, I was just having to work harder to get the same sense of achievement. Whole mornings were spent sitting on rocky outcrops gazing out to sea while whale watching and trying to rationalise things. The running had assumed a robotic quality that had the potential to obscure everything around me. Balance was difficult to find and mileage remained the priority. I told myself sternly that the negativity of niggling injuries was causing me to question my motivation, and that these were the moments in which rising above feelings mattered most. The focus had to remain pure and never waver.

While staying at a charming backpackers' lodgings in Arcata I was able to enjoy a bath with lavender oil and thanks to a kind donation was able to splash out at a sushi bar. Unfortunately, on leaving the next morning the local police were not quite as accommodating, pulling me over and demanding I refrain from running on the asphalt and move onto the grass verge. I complied for about five metres until they had driven around the corner.

Later that afternoon while the Californian sun was beating down, I developed a real hankering for ice cream but there was very little around except a small business that built greenhouses. 'You don't happen to know where I could buy ice cream around here?' I asked a large friendly-looking chap with a goatee and baseball cap. 'Sorry dude, nothing around here,' he replied but as I turned to the door he added, 'I have a cold beer if you want one?' I stopped dead in my tracks, swivelled and seconds later was sipping a nice cold beer with Jon.

We chatted for ages. Perhaps I was a distraction from his day and for me, he was vital human contact. After thanking him and saying our goodbyes, who should charge up in his car about an hour or so later, but Jon? The pretext was to check I was alright, and he was clearly startled by how far I had got. He also wanted to get a selfie as he thought he may be able to drum up some local newspaper interest.

As the sun began to set, I heard a familiar voice behind me. Allan and I decided to finish the day together and find somewhere to camp. We arrived in Scotia, a small community built around a lumber yard, but there were no camping spots to be found. The town's hotel was tempting, but way out of our price range. Over a cold beer at the bar we learnt that the next campsite was in Stafford, about 5km further down the valley. Having already done 60km, it was hard to find the energy, but we pushed on.

After we had set up camp and food was on the go, a car pulled up and out jumped Jon. Earlier, I'd told him about my Spot GPS tracker, which my brother had given me to make sure I could always be found. Having used it to locate me, he'd thought it would be fun to join us and pulled out a six-pack of beer and the biggest bud of marijuana I had ever seen. He explained that we were entering the Emerald Triangle, the largest marijuana-producing area of the United States—hence the demand for greenhouses! It would have been fun to have said that we all kicked back and shared a massive blunt, but I excused myself and stuck to the beers. That evening yielded another surprise, when Jon told me that the local TV station was going to stop by the next morning for an interview. Halfway through filming, the interview was cut short due to a police shooting that was reported over the scanner. Welcome to America!

Over the next couple of days, the scenery got impressive while winding through the Avenue of the Giants, which carves its way through the Humboldt Redwoods State Park. It was there that I met a 70-year-old cyclist called Jim on his way across America to visit his daughter.

'I don't want to be like all the other people my age who have resigned themselves to a life of sofas and TV in a condo,' he explained, before reaching into his pocket and giving me $20. 'Please take this and buy yourself something nice. To repay me, later down the road, you give $20 to someone to enjoy, and hopefully, this will continue.' I took the money and vowed to spend it on something delicious and did indeed pass the money on, a few months later in Mexico.

Lots of little distractions broke up the next few days. On a diversion through a village called Redway for a radio interview that Jon had set up, I came across a proper chain gang. Ten prisoners, dressed in orange overalls were working along the verge of the road, watched by policemen with shotguns. I kept my distance as I was ushered on by the police under strict instructions not to interact with anyone. Following the interview, Jon and I met up for the last time, after he had booked and paid for a night at a local campsite. Over a great dinner together, we vowed to keep in touch. Sadly, I haven't been able to track down Jon since I returned from my adventure.

With the weather heating up, the midday sun was taking its toll. This necessitated taking lots of breaks whenever there was shade—under trees, bridges—anywhere to recharge the energy which the sun had siphoned from me.

It was hard to ignore blaring music as I passed the Redwoods River RV Park or the makeshift sign against the gatepost that read: Soul Camp. But by this point I was becoming an automaton, so ran on.

A kilometre further down the road I was wrestling with doubt. What was I doing? Wasn't I fed up with not exploring enough on this trip and becoming a bit too focused on distance? Might this Soul Camp offer some respite? Doubling back, I approached a desk at the gate and asked what was going on.

'Soul Camp is an event that is driven by the soul, run by genuine people from all walks of life who possess the same passions. Everyone's main goal is to be happy, have fun, and live drama- and argument-free. Basically, we are a bunch of DJs and artists from San Francisco and around, and each year we get together, take over this park and have a party,' the attendant informed me.

'Is it open to everyone? Can I join?' I asked and was told yes, in return for $30. My hopes for a break were fast unravelling. Having run out of cash, I'd been relying on my bank card. The only shop on site didn't do cashback, leaving me in a frustrating predicament.

Perhaps I shouldn't have worried as I clearly didn't resemble the typical Soul Camper. Looking me up and down, the guy finally asked what I was doing. On explaining my adventure, he laughed, 'If that is the case, then why are we worrying about money, make yourself at home!'

With no idea what to expect, I busied myself setting up my tent and stocking up on provisions from the small camp shop. The positivity was contagious, and everywhere I looked people were smiling, happy and super friendly. Any feelings of unease rapidly dispersed. That evening I found myself dancing to trance music in the trees with strobe lights and beers. A Mexican lady of short stature approached me, holding up a container of gummy bears. 'They have been soaked in vodka', she grinned, encouraging me to dig in. The music played until the sun rose.

After dancing among the trees until dawn, a day off seemed like a good idea, especially after covering nearly 250km over the last six days. What better way to relax than a pool party and it wasn't long before I was being offered more snacks by my new Mexican friend and her husband.

'Want a hash brownie?' she asked. Not really being one for drugs, it took a little persuasion but I was here to experiment and experience more, so was soon nibbling away. When they asked me to join them for lunch, I enthusiastically accepted. By the time we sat down I was predictably less communicative; all I remember was sitting at the table either shovelling huge amounts of food into my mouth or sitting slouched in a chair staring into the distance. Thankfully, an afternoon nap brought me back around and it wasn't long before I was dancing to a mixture of reggae, funk and Latin music played by a local band called The Bayonics. Afternoon became evening, and a party kicked off with more beers and DJs. Despite having to run an ultra-marathon distance the next day, I went with the flow and relished this short escape from the daunting prospect of the road ahead.

Somehow, the next morning wasn't too bad. I had let my hair down and partied for two nights. While it wasn't the healthiest form of recovery, the interaction with people lifted my spirits and sharing my stories with them reminded me of my mission. I left with a spring in my step and more than 55km passed trouble-free, including long, steep hills before leaving Highway 101 and arriving at the breathtaking State Route 1. That night, camping on the cliffs next to Cottaneva Creek, I felt rejuvenated with a hunger to power on.

Being on the iconic Highway 1 was a real game-changer. With the Pacific Ocean on my right, every day was therapeutic and the increase in hills imperceptible amid the challenge and the captivating beauty. The adventure was becoming enjoyable once again, and as my running ability increased and injuries subsided, I was allowing myself to be more spontaneous. Even being kicked out of a Starbucks because they didn't want my stroller inside didn't get to me. Instead, I cheerfully relocated to a smokehouse and devoured a pulled pork sandwich and numerous cans of sugary soda. Another 10km brought me to Russian Gulch, a small bay, where I set up camp next to Caspar Beach. Unbelievably, after 39 days of running with the Pacific as my companion, this was the first time I had actually ventured into the water and any notion that the Californian ocean was going to be warm was dispelled as the icy water washed over me.

Back at the campsite a lady enquired what I was doing, an increasingly common event by now. Curiosity is an innate human characteristic, and although I felt no different, outwardly my increasingly dishevelled appearance paired with running kit and a baby stroller was clearly provoking interest among the people I encountered. When I explained, she smiled and just said, 'You are not going to regret this, you are going to be one of the lucky people who will look back on life and realise you didn't waste it.' And with that, she handed me some coins for a hot shower.

At Mendocino there was a café stuffed full of cyclists who wryly explained it was nicknamed 'Spendocino', due to the high price of everything, including the coffee. Biting the bullet and in preparation for a big push that day, the full breakfast was a welcome break from the porridge sachets I was becoming used to. Leaving the café, the running was hard along the more exposed, windswept coastal landscape but the new fire in my belly propelled me forward. I arrived at the Manchester KOA campsite to find several of the cyclists I had met over the last few days already there and we spent the evening with cold beers, sharing tales from the road and soaking sore limbs in a hot tub.

California's coastal weather proved ever changeable, and the next day's heavy rain meant running was off the cards. By the end of the day everyone's tents had flooded and all but two of us had relocated to the more expensive cabins. Just me and a girl who was bikepacking toughed it out. To a Northern European, it was mystifying that a little rain seemed to cause such upheaval but the staff at the camp were extraordinarily generous and opened a large shed where they promptly served hot meals to all campers. It was how I imagine disaster relief, albeit on a much smaller scale. Drying myself even for a second was nice, but the camaraderie we felt was the real treat.

The rain persisted intermittently throughout the next day and that evening I found myself in a small, damp campsite near the town of Gualala. While chatting to a couple pitched next to me, they pointed at a tent and asked if it was mine. On confirming it was, they told me that I should probably get back there as it was being ransacked by raccoons! Until this point wildlife hadn't caused any issues, so the havoc those little bastards could inflict was a shock. Just in time, I managed to temporarily scare them away and save my food, which was now strewn around my pitch. On closer inspection, it seemed they had a recipe in mind as they had focused their efforts on bagels, peanut butter and jam! Even the guidance of Allan and the others wasn't enough as when I found the food boxes, it was clear the raccoons had already eaten through the damp wood, rendering them useless. My only option was to bag up all my food to store in my tent and hope for the best. That night was harrowing, with repeated raccoon attacks where they would physically run into the tent. Their persistence contributed to my foul, unrested mood on waking, which prompted me to pack up early and escape as quickly as possible.

The road always brings positives to lift your mood and, although I never received the gift, I heard from cyclists that someone had left a care package on the side of the road for me with a little sign saying, 'For the runner'. Sadly, I either missed it or someone had helped themselves, but regardless, this act of kindness put the previous night behind me as I turned my focus towards San Francisco.

San Francisco was pulling me in. The last big city had been Seattle some 1,400km ago and while I preferred to avoid cities, San Francisco had some magic attraction.

My daily mileage was creeping up again and covering distance was becoming a manic preoccupation. This prompted my father to get in contact to tell me to slow down. I am very lucky to have supportive parents who have been there for me my whole life and have given me every opportunity I could have wished for, though whether I have taken full advantage of that is another question – something I pondered a lot while running. One of the things I noticed about this adventure was the new closeness that developed between me and my parents, notably with my father. While in London, my interactions with him weren't that deep but something had changed. I can only think that he recognised my passion and that it resonated with him. While contemplating this adventure, I'd worried that it would disappoint my parents, to which my father pointed out that he had never pushed his children in any direction, other than what motivated us and would help make a living. To back up his advice he sent me the song Feeling Groovy by Simon & Garfunkel, who tell us to slow down.

Taking my father's advice and after 46 days on the road, I decided to take an extended break around San Francisco. My first pause was in the beautiful Sausalito, just north of the Golden Gate Bridge. The main reason for stopping here was to take advantage of some much-needed hospitality from Vicky, a friend who had kindly offered some normality for a couple of nights. I arrived at lunchtime with some time to kill. Sitting against a wall to make some calls, one of the locals thought I was homeless and walked up to me and benevolently dropped $50 into my lap. Another approached me to ask if he could help and before long, we're sitting outside a bar drinking beer. Conversation turned to living a life without regrets and I start to realise that when people see me on this adventure, it makes them reflect on the opportunities or adventures they may have passed up in their lives.

Spending time with Vicky and her then-fiancé, (now husband) Peter, was great but normality evaded me after spending so much time on the road alone and free. Things like dinner parties just seemed so alien and being the target of most of the questions was a little intimidating.

The main reason for stopping north of San Francisco was to cross the Golden Gate Bridge in the morning for its full impact. I'd been building this immense structure up in my mind and couldn't wait to see it. CALM, the male suicide charity, was a major beneficiary of my run, so conversations had frequently moved to this subject and the more I learnt, the more it compelled me forward. Sadly, the Golden Gate Bridge has the second-highest suicide toll of any bridge in the world with over 1,600 known suicides, second only to the Nanjing Yangtze River Bridge in China. Arriving here early then was crucial, without needing to rush or be anywhere else, to just reflect and remember the 12 men a day who take their lives in the UK.

CHAPTER 6 - AN ENGAGEMENT AND FORREST GUMP

DISTANCE TO BUENOS AIRES: 15,220KM

My first experience of San Francisco was a misunderstanding at the first hostel I tried to stay at, resulting in my being kicked out. Upon arrival at the Fisherman's Wharf Hostel with camping gas and newly purchased cold cider, I was promptly asked to leave for breaking too many regulations. Slightly disheartened, a journey to the centre of town sorted me out and before long I was enjoying everything on offer. This included booking myself a tour of Alcatraz where Ai WeiWei was holding a special exhibition featuring a giant paper dragon, porcelain petals and six large carpets of Lego blocks, depicting more than 175 prisoners of conscience. I kicked back and relaxed at Ghirardelli Ice Cream and Chocolate Shop, as well tasting the best Irish coffee I've ever had at the Buena Vista. At In and Out Burger, a bride ran in, quickly followed by her husband and bridesmaids, to order burgers and shakes to the applause of the customers! The sights and sounds of San Francisco could not have been more welcome and I took it all in as if seeing urban life for the first time.

That evening, I caught up with Emilien and Helen, two Dutch cyclists I'd sheltered from the rain with back in Manchester. This would be the end of their journey, and it was heartening to share that with them on a night out. And what could end this amusing evening better than sneaking into the Hyatt Hotel with one of their friends, who insisted I would have a more comfortable night there than in a six-bed dormitory?

However nice this interlude had been, the call of the road was strong, and I was becoming restless strolling around the streets and parks with neither purpose nor motivation. After two days' rest I headed out of the main city, the skyscrapers and apartment blocks falling away to be replaced by a sense of being at home, alone, and in search of new experiences.

It wasn't long before the journey delivered another colourful character. After running 67km from Half Moon Bay, I struggled to find a suitable camping spot in the small town of Davenport. Unsettled weather conditions and so many people meant the beaches weren't really an option, and I found myself running back and forth in vain. A passer-by spotted my confusion and asked if they could help. 'You need to meet Joe Ray,' they said, 'He'll be able to sort you out.'

Joe was a windsurfer and as it turned out was a sort of local legend in Davenport. On hearing my predicament, he immediately offered his workshop as a place to crash. It was a huge shed with walls adorned with fishing rods, drums, surfboards and antler plaques. I set up camp on the huge workbench where he mended windsurfing sails. His hospitality didn't end there and soon he had me in the local restaurant with the friends that had introduced us. The next morning, we shared breakfast and chatted more. Joe was one of the most laid-back men I have ever met and he regaled me with stories of working in LA, windsurfing big waves and pursuing a life filled with passion and excitement. As I sat there, unbeknownst to him, his enthusiasm was stoking my desire to make this change in my own life a permanent course alteration, rather than a mere diversion. It still feels strange that so many meetings of kindred spirits are over so quickly and without any real evidence that they happened at all.

In Santa Cruz, a girl called Maryam offered me a sofa in a college house at the university. I am not going to lie, in my mind I pictured staying in a sorority house with cheerleaders like so many films depict, but in reality it was a lovely house with a great group of girls who went above and beyond to make me feel welcome. They persuaded me to take a day off and go to university to attend some of their classes. The first of which was an interactive class where we all had to prepare material and share it, which I found rather daunting. The afternoon included a boxercise class, which after 1,950km may have been a strange decision! There was one strange part of this stay and that was when I was packing to leave and found maggots infesting where my bags were. Luckily there were none inside my bags and I checked all my equipment and couldn't find any source and had never experienced this before (and never did again). Rather than confronting this issue, I made a sharp exit and put it down to university life! They might remember this very differently and be cursing that they ever invited a traveller into their house. Hopefully we can all smile about it now.

The Californian coast continued to be crammed with beauty and kindness, and I had a real sense of being at one with my adventure. I understood my systems, trusted my ability and was slowly allowing myself to enjoy the experiences as they were presented to me.

No specific moments define this next stretch but rather a compilation of small moments that helped sustained the momentum. On one occasion, after running through acres of fruit farms with the intoxicating smell of sweet strawberries filling my lungs, there was nowhere to camp. As was becoming the pattern, when help was most needed, it seemed to materialise. A girl appeared out of the ether and offered her back garden as a camping spot. She was young and home alone so understandably wary of having a man in the house. However, she joined me in the garden and the more we talked the more her anxiety eased and she ended up persuading me to have a hot shower and we enjoyed a delicious meal she prepared.

Monterey introduced me to Trader Joe's, an American supermarket, which dramatically raised the quality of my provisions, which was a godsend given the road ahead. My diet had become quite unimaginative with easy foods like porridge, bagels and pasta making up the bulk of my carbs. This was now supplemented with treats like chocolate covered gingers, or popcorn that I cooked in my

Trangia stove with coconut oil and ground pepper. While in a café toilet, a kind passer-by left a selection of cereal bars on my stroller, which gave me an energy boost as the road began to soar. 'Roads don't soar…' you may say, but entering the Carmel Highlands and the Andrew Morela State Park did feel like flying. The road gripped the edges of the cliffs and I felt lighter than air, especially when breaking through the clouds and ocean mist to look down on a carpet of white stretched out below. Whenever the Pacific was visible, my rests would be spent watching the dolphins below and soaking up the sun. I always felt that my experience of this extraordinary beauty felt so much more special compared to that of the car passengers who sped by, trying to snap a photo through an open window.

The highs and lows were not just topographic. One night I arrived at a campsite only to be turned away because there was no space, forcing me back onto the precarious highway to continue running in the dark, something I tried to avoid. But this was followed by a much better experience at the next campsite where a group in a campervan invited me over for an evening of salmon, drinks and laughs around a campfire. As generous as the people I met were, a life on the road left me open to other hazards. For example, falling prey to a passing seagull that swooped down on my plate and stole my lunch, much to the amusement of those around. Fortune rebalanced things again, when a British tourist driving a Ferrari pulled over and restocked my stroller with energy bars and cold drinks. It turned out he had spotted the Union Jack flag as he drove by and felt compelled to help out.

By San Simeon, I acknowledge that my personal hygiene had slipped. It had been six days since my last shower and I'd covered over 200km wearing the same clothes to run and sleep in. This became apparent on entering a shop and being distracted by the smell, which I ashamedly had to accept was emanating from me. To make things worse, the temperature outside was soaring and it was becoming increasingly hard to find water to drink, let alone wash in.

Luckily, Morro Bay provided a quaint respite from the road, including a shower, a couple of calzone pizzas and more help from fellow travellers. Ben and Paula were driving a campervan and we had crossed paths the day before. During the nights they took all my electronics and would charge them up for me as well as providing snacks and drinks. In essence it was like having a mini-support team for a couple of days.

Santa Barbara was the next major marker on the map and the closer I got the clearer it became that road restrictions on Highway 1 were going to prevent me from taking my planned route. There was no choice but to cut east and take the more arduous road over the hills through the insanely beautiful wine region along Foxen Canyon to Los Olivos. A group of farmworkers had congregated outside a grocery store and invited me to join them. We sat chatting about everything and nothing but whiled away the balmy evening with the beers we plucked straight from the store fridge. When it came to finding a place to camp, they secured me a spot in nearby wasteland that had been used to grow mushrooms! Of all of the USA, these were among my fondest social memories.

Santa Barbara marked the beginning of the end of my USA stage, as my journey curved round to Mexico, passing through Malibu, Los Angeles and San Diego. Descending the hills into the town itself, I allowed myself to believe that running the length of the USA was very much within reach. An adventure of this magnitude was always going to be a huge physical challenge, but the mental aspect was undeniably the bigger obstacle. Not having undertaken a project like this before, I needed a strategy to deal with it, so from the outset I'd defined markers of success along the route, and running the length of the USA was the first of these. Rather than focus on the finish line, I would target the next marker and allow myself to feel more frequent feelings of achievement. On a personal level, I was potentially about to accomplish something that, for the first time in my life, I felt I could be proud of, something I had conceived, planned and executed successfully.

Near the high point of the San Marcos Pass I received a call from my little sister. 'We've got some news...' Melissa said. I knew what was to follow before she could say it because her boyfriend had taken me aside at one of my leaving drinks and let me know his plan.

However, that had been at the beginning of August and over two months had since passed, and to be honest I'd started to worry that he may have changed his mind.

'Shep has asked me to marry him and I have of course said yes!' I don't think it was any secret she was keen to settle down and start a family and I couldn't have been happier. It is relatively easy for me to distance myself from people or social groups anyway, but when something special happens I do feel a pang of isolation. Vowing to celebrate in Santa Barbara and with a new spring in my step, I relished the descent to the ocean. At a smart bar on the main street, I raised a glass of prosecco to my sister and her fiancé, then selfishly wondered when the wedding would be as I wasn't scheduled to be back for over a year.

Everything in Santa Barbara was way out of my budget, so I resorted to Warmshowers and left the following message at about 4pm.

'Sorry for the late notice but being a runner it is hard to gauge arrivals! I am running from Vancouver to Buenos Aires and am looking for a place in Santa Barbara tonight. I know I am being optimistic!'

After a tense couple of hours, a chap called 'KG' made contact. He said he could help but I would have to come to the marina, as he lived on a dive boat. Somewhat dubious, I made my way down to the oceanfront and was duly swept into the larger-than-life world of KG and his fellow divers. After my stroller was heaved aboard, I was introduced to a table of dive instructors playing drinking games and, needless to say, it took very little persuasion to get me involved. The night rolled on and, leaving the boat, we made our way to small bars where pints of beer were followed by whiskey chasers and punctuated with games of beer pong and table football. I hadn't let my hair down like this since Soul Camp a month earlier and being included filled an unspoken need. While the fun went on into the small hours, the re-appearance of my old friend the hangover was less welcome.

Note from KG: 'Jamie is an amazing guy. He showed up to my boat while we were just finishing cleaning after a trip and were a bit inebriated and in a party mood. Even after a long day of running, he joined in our festivities and we had a great time. I hope your journey goes well.'

Unsurprisingly, running to Malibu that day was slow, with long stints in cafés trying to find the energy to get going and eliminate the raging and stubbornly persistent headache. On stopping for lunch in a small beachside restaurant in Carpinteria, I went to pay the bill, only to find that another customer had already settled it without any fuss, just a silent act of kindness. Various cyclists joined me for short stints and showed me the best routes to help avoid traffic. The highlights of this stretch were simple; another opportunity to swim in the ocean at Sycamore Canyon, running by the Point Mugu Missile Park that had a real F14 Tomcat on display, and buying fresh strawberries on the side of the road.

After 350km in seven days, Malibu was my last stop before Santa Monica and I had hoped for some celebrity spotting while devouring pizza at the famous D'Amores restaurant or outside Pacific Coast Greens supermarket but alas, it was not to be.

The main pull to Santa Monica was an old university friend called Dom who worked for Headspace, the meditation app, and who had kindly agreed to receive a couple of packages at the company's head office in Venice Beach. Despite it being my day off, I jogged there with my stroller to give them the full adventurer effect, although perhaps I was already displaying signs of an unhealthy attachment. On collecting my kit, it meant the world to have lunch with someone I actually knew and could feel relaxed around.

The rest of my time in Santa Monica was spent sorting out my equipment, resupplying, relaxing and getting drunk with fellow travellers. I'd missed cinema while on the road so tried to take advantage of this as much as possible. The first night it was Gone Girl and then on the second, Fury, which I watched with a girl I'd met in the hostel. Our little excursion led to drinks and silliness that reminded me of the life I'd left behind and while enjoyable, made me miss the simplicity of being on the road. A theme was beginning to emerge in that after a while on the road I would crave company and socialising, but it very quickly would lose its appeal and I'd simply want to continue running.

Notwithstanding the urge to get moving again, I had a house call to make in Redondo Beach, just 25km south. As there were so few occasions to see familiar faces, I took advantage of them whenever I could.

Despite not having seen Fiona for over 15 years, it was amazingly easy to slip back into friendship. The family environment was comforting and I could relax a little more than I had been accustomed to. It was also a perfect opportunity to take advantage of her husband's tools and see to some much-needed repairs.

San Diego was only a few days south and spending so much time in built-up areas without camping as much was playing on my mind. Craving the basic routine of pitching a tent and cooking for myself, I tried to push the mileage back up and make it somewhere more rustic. Speeding through Huntington Beach and Newport, I finally reached Crystal Cove. I hadn't appreciated how much of the land was private or how expensive campgrounds were. Typically, state parks are the cheapest option but not this time. The sun was setting as I arrived at the Moro Campground and proceeded to the office, where the wardens informed me that a simple pitch was going to cost $50. For someone on a tight budget and without sufficient money in the bank to complete my adventure, I pleaded for flexibility. Would it help that I didn't need all the facilities and would be gone in the morning? Unfortunately, they weren't budging so it was either pay the fee or leave the park. I couldn't justify the cost but it was getting dark so options were limited. The wardens huddled around chatting and I was starting to think they were being petty but in fact I was completely misreading the situation. It transpired that, rather than kicking me out, they were doing a whip-round to find the money to buy me a plot for the evening. Bowled over by their generosity and after hugging them all, I pitched my tent on an idyllic plot overlooking the ocean, counting my blessings. My neighbours were a lovely British couple called Doug and Lynn, and their children, who immediately took me under their wing with beer, food and brilliant conversation. They too were people who had looked at life, disputed the norm, then upped sticks and now lived on their terms. We remain friends to this day.

Running in Southern California was blissful as it had a nice balance between being in nature and a lot of visual stimulus from small villages and the beautiful coastline. That said, camping continued to be limited, which dictated each day's running. Wild camping is one thing but trying to stealth camp day after day in populated areas is a huge mental drain and one I desperately tried to avoid, especially in California where the rules are strict and people have guns! One night I detoured off the main road in search of the San Mateo campsite, which due to it being late October, was not too busy. Campsite wardens are more often than not retired people who monitor the park in exchange for free parking of their RV. I made myself known to the man in charge at San Mateo and he suggested that, rather than paying for a pitch, I just camp behind his RV. That night we sat outside drinking beer while he regaled me with stories of his life that he had turned into a book, Dead End Street by Dean McCormick. Apparently, he was a surfer whose life turned to addiction and drug trafficking, which took him as far as Asia. His life had been quite dark at times but he appeared to be making the best of things. Like so many of the people I met in North America, he was open and kind, despite his own struggles along the way.

The route to Carlsbad was a little more exciting with a trail down the side of Route 5 before veering inland towards Camp Pendleton, the main West Coast base of the United States Marine Corps and one of the largest in the US. At first glance, the route looked like a dead-end with a checkpoint manned by two marines, who were casually eating takeaway pizza. One of the guards approached me and demanded my identification, which I duly provided. A rather lacklustre search of my stroller ensued, and minutes later I was running through the actual base alongside tank tracks and soldiers on drill. The road then meandered through the barracks before emerging back onto the main road near Oceanside. It was a bizarre diversion but one that provided plenty to take my mind off the running.

On approaching Carlsbad, an opportunity for some luxury arose. The daughter of Ben and Paula, who had been so good to me in San Simeon and Morro Bay, had written me the following email:

'My parents Ben & Paula were very touched meeting you along their travels. They have not been on vacation in over fifteen years and rented a motorhome. One of their best memories thus far was meeting you! Thank you for that and they say hello. If you are in Carlsbad, CA hit me up...'

On which basis, as I approached, I replied to her email with:

'Hi, if you happen to know any good cheap campsites that would be invaluable intel!!!'

The reply came back, 'I called the Carlsbad State Campground Park Ranger and they have plenty of availability. I also have rooms at the hotel if you would like a nice shower, bed, etc.'

It transpired that she was the concierge for the Hilton Hotel, and despite my desire to camp again, I couldn't pass up this opportunity! Although I arrived at the Hilton feeling so out of place, I was greeted with unbelievable kindness and shown to a room with a door out onto the pool. On my bed was a fruit basket and even some shaving cream and a razor. I certainly was in luxury and wholeheartedly appreciated it, but it felt very apparent that I was the outsider among the other guests and when sitting in the bar I felt even more alone than I did when in my tent. There was nothing else for it but to withdraw to my room and, rather than taking advantage of everything on offer, set about the usual tasks of cleaning my kit, fixing issues and planning my next steps.

San Diego was a big milestone before my adventure would continue south into Mexico, so I decided to make a more significant stop in Ocean Beach just north of the main city and celebrate my 35th birthday. Described as a vibrant, bohemian neighbourhood, with a classic SoCal beach vibe, it proved the ideal place to recover, reflect and prepare for what lay ahead.

First, I set aside some time for some touristy things, such as visiting the USS Midway and the lively downtown Gaslamp Quarter. I was also enjoying wandering the streets, appreciating still being in a country where I could understand everything that was said or written on signs. It felt important to visit restaurants and dine on all the foods I thought might become scarce south of the border.

San Diego was also the scene of a fleeting romance. To set the scene, I was staying in the Ocean Beach hostel on the main street up from the beach. It was brightly coloured with purple fences and had a huge peace sign on the roof and a very communal vibe. On the first night in the hostel I got chatting to a few people over drinks, including a South African girl called Jamey who was in San Diego looking for a job on a luxury super yacht. It was Halloween the next night and Americans sure like to celebrate this holiday. Everyone was talking about their plans and what they were going to wear, but I wasn't committing to anything as I just wanted to decompress.

The next day Jamey and I met again, and immediately hit it off. She was beautiful, free-spirited and fun and there was an immediate spark. We hung out for most of the night and, when things died down in the hostel, we made our way over to the local bar for more drinks, dancing and then finally, at the stroke of midnight I got a birthday kiss. Unfortunately, Jamey was heading off the next day, but we agreed to spend my last evening in San Diego together and have a late birthday meal. Divulging no details, it felt really good to have a proper physical attraction and emotional interaction with someone after so long on my own only really experiencing short, passing connections.

Alone once again the following day, I had time to myself to reflect on the eventful start to my adventure. I had all but run the length of the United States, a total of over 2,800km in just 77 days, averaging over 35km a day, or 44km if you exclude rest days. I had reached my first success marker and as a result, my expectations were high for how that might make me feel. I went for a walk to think about how everything had affected me and tried to analyse the benefits of my choices and acknowledge my mistakes. If you've ever been to Ocean Beach then you'll know about the Ocean Beach Pier that juts out nearly 2,000 feet into the Pacific Ocean. It provided the perfect place for reflection.

I thought, or maybe hoped, that being there might have felt more magical, having pushed myself so hard and overcome so many obstacles, big and small. Now, I found myself driven by a desire to eke out as much from myself as I could. And while I was, of course, proud of my achievements at this point, it was also clear that I was becoming immune to the challenges presented by running through the USA, a developed country with English-speaking people and very little danger. Was there even a touch of boredom creeping in at not being able to challenge my boundaries to their absolute limit because I had become comfortable with everything? Of course there were tinges of emptiness at this thought, so my practical side had to turn this into a positive, especially as my adventure was about to ramp up quite a bit and the challenges would inevitably become more demanding. Looking back, I'm not sure if my mind was prepared, but as I stood in San Diego recalling everything I had loved about the USA, it felt like the right moment to leave it behind and take on the next chapter of this adventure. The USA had served its purpose perfectly, to prepare me physically and mentally, and equip me with the necessary skills and confidence to face more adversity.

The last thing I did in San Diego was to change the course of my adventure. Originally, I was going to cross the border into Tijuana, then head east across the Mexican mainland, veer south towards Hermosillo and on to Guatemala and beyond. However, since meeting Jon up in Humboldt County, the suggestion of running Baja California had been at the back of my mind. Jon had a friend in San Diego, who also had a place in Baja, and he agreed to meet me to discuss the options.

I met Justin at the South Beach Bar and Grille and instantly liked him. He was a tall, good-looking surfer dude with a shaved head and a scar across his hairline. He was also just a nice guy who reminded me of friends from home, and even though I didn't know him, it felt like hanging out with an old friend! His impressive knowledge of Mexico also came in handy.

Entering Mexico was increasingly on my mind as I neared the border. To me, there would be no issue with running there but it was hard not to be affected by the constant warnings and horror stories that the many Americans I had met shared with me. The most common being the recent discovery of human remains found in two plastic bags near a beach in La Majahua, southwestern Mexico, a place I would visit two months later. They belonged to Harry Devert, a New Yorker who left his job as a trader in finance for a transcontinental motorcycle journey from the United States to Latin America. For obvious reasons, this had got a lot of news coverage in America and as a result, everyone seemed to fear travelling south. Raising my concerns with Justin, he thankfully put my fears to rest, reporting that Mexico and especially Baja was a fantastic place, whose people were kind and had a great sense of adventure. It was soon confirmed that my run would now stick to Baja California and my next checkpoint would be La Paz, some 1,500km away.

My last afternoon in San Diego was spent being shown around by a girl called Kimberly, who happened to be the ex-girlfriend of Matt, the hitchhiker who had sourced our beers in the Redwoods Forrest. She in turn took me on a beer tour around the area just outside San Diego and introduced me to Matthew, who was set to be my host in Tijuana the next evening.

As I lay in bed on my last night in the USA, I thought back to my first interaction with Americans in that supermarket in Blaine and the check-out lady who reacted so positively to my adventure. Admittedly, I had been sceptical about how I would get on with the American people. My preconception was that they might grate with my personality but after spending nearly three months encountering the most amazing people and receiving unquestioning kindness and support, I had to concede that America really is an amazing place. Dreams can come true there, and just maybe, the Americans' positive reaction to what I was setting out to achieve gave me just the start that I needed and the confidence to keep forging on.

PART TWO – CENTRAL AMERICA

CHAPTER 7 - A MEXICAN WELCOME

DISTANCE TO BUENOS AIRES: 14,193KM

Having spent more time in San Diego than anywhere else since my adventure began, and with the delicious anticipation of the greater challenges ahead, I was keen to get back on the road. Tijuana lay just 40km south and having had a few days off, I had to pace myself to avoid more injuries. Revisiting places along the waterfront, my first issue was constant punctures, with three happening simultaneously. These turned out to be due to goatshead thorns, or puncturevines as they are also known. Little did I know that these little buggers were going to be a constant nuisance throughout Baja. Two policemen pulled me over and refilled my water bottles and reiterated everyone's fears about me travelling south to Mexico. However, my excitement suppressed all these potential fears and nothing would deter me. The only remaining mission to fulfil before leaving the US was to buy new running shoes before I disappeared into the unknown.

The expedition began using the old running shoes I'd had in the UK because I didn't have the money to invest in new ones. My system thereafter was pretty simple. One pair to run in and one spare, swapping every few days to rest the shoes and give my legs a slightly different running motion. I had managed to scrape together enough money to buy new shoes in Lincoln City, Oregon, and San Francisco and had relied on finding outlet stores to provide cheap options. Knowing it was going to be harder to buy genuine running shoes in Mexico, I was glad to source a new pair.

At rush hour, the border was hectic because many Mexicans were returning south after a day's work. This was my first crossing into a foreign-language country and with a stroller in tow, with hindsight, doing this in daylight would have helped. The actual process was relatively easy, if a little frantic. Once again, an imaginary line divided two very different worlds yet crossing it into the streets of Tijuana was very real. It was a huge relief, and I felt neither scared nor intimidated, just bizarrely relaxed and full of excitement for what lay ahead. To me, Tijuana seemed more welcoming than the US, where everything is very orderly, clean and precise. Here people were crowding around parks, selling street food and playing loud music. The tight constraints north of the border did not apply here and that seemed to be reflected in the people.

Having somewhere to sleep that night and Matt, whom I had met in San Diego, to introduce me to Mexico definitely helped. Not long after arriving at Matt's flat we were heading out into the hustle and bustle of the local area. Matt was keen to show me his Tijuana, not the Tijuana you hear sketchy and ominous stories about from their northern neighbour. Our first stop was a hole-in-the-wall taco restaurant which reportedly serves the best tacos in the city, and so began my obsession with real Mexican food. Over a large IPA in a craft beer pub, Matt interviewed me for the San Diego Reader, where he was a journalist. The night meandered through small dive bars, the red-light district and even the transvestite area of the town. It was quite the introduction! The streets were alive with people, including Americans who had made it down here to take advantage of things they couldn't do back home. With no sense of risk or danger here, I ended my first evening feeling more at ease about what lay ahead.

The 4th November was my first full running day in Mexico, and to say I was raring to go is an understatement. However, despite lots of time to relax the night before, I hadn't planned which route out of Tijuana was the safest option. Matt had warned of the gangland areas along the way and said he was going to give me an alternative, but when the morning arrived he was in too much of a rush to get to work and hesitantly told me to take my original route and tried to assure me that it would be fine. As unsettling as this was, especially with all the warnings since Seattle, I didn't have much choice except to head off with a positive attitude and trust in people.

My chosen route was a bit of a gamble as it cut west and relied on being allowed to run down the hard shoulder of a motorway. This of course would ensure contact with the local police, or Federales as they were known there. My main concern was whether they would allow me to run or just send me back. Then there was the risk of being extorted. The main benefit of the highway was its smooth running surface, so I decided to prioritise that. Leaving Tijuana, the road was more treacherous than those I had been used to, with more potholes and no hard shoulder, not to mention the motorists' questionable driving skills. The surrounding area all felt very alien, if exhilarating at the same time. I'd been waiting for this moment for so long and to have the hairs on the back of my neck standing alert was adding to the whole experience. Adventuring in a new land was everything I wanted.

As predicted, a group of Federales was waiting for me at the toll road. It was my first interaction with the much-maligned police force but to my relief, everything worked out perfectly and with minimal discussion. In broken Spanish I explained that I was running to Argentina and they agreed that the freeway would be safer. This nice, uncomplicated start in Mexico helped with the mental burden of venturing into the unknown. However, despite the beautiful, roomy hard shoulder and light traffic, a goatshead thorn puncture soon happened, necessitating a stop at a small shop. Here and in the days that followed, I was taken aback by the interest from everyone I met—though admittedly you don't often see runners on the freeway—and how helpful and friendly they were.

My endpoint was kilometre 38 on the highway, where Justin had arranged a place for me to camp. Here everything was sandy, desert-like and scorched, with palm trees and unfinished houses along the road. It might sound bleak, but being away from everything familiar was relaxing. I couldn't understand any of the Spanish-language signs but the huge US influence was unmissable, with massive billboards stretching the names of Walmart and McDonald's across the horizon.

Bringing the stroller to a halt, Roberto Cortez was there to greet me, a friend of Justin's who said he'd put me up for the night. I was a little surprised when he showed me onto the deck of someone else's property and pointed out where to hang my hammock. This was completely alien. I was meeting someone who knew someone who knew someone I had met in northern California, and he was showing me to someone's house who wasn't even there! To get through this adventure, I was going to have to trust people, so I tied my hammock up on the deck overlooking the coast and surrendered to what the future held.

After a couple of marathon-distance days, I pulled into La Fonda, a small surfing village snuggled between the highway and the ocean. Having surveyed the ocean for so long, it seemed time to give surfing a go. I booked into a rather curious hostel run by an old chap who lived there with his, I want to say, daughter. It felt more like a home where people dossed for a couple of days and seemed popular with surfers. Over dinner, I managed to commission the services of a British guy who would give me a surf lesson. Swapping the monotonous tarmac for the cooling waves of the ocean was a welcome distraction, and while I couldn't quite re-enact scenes from Point Break, hopefully I did all right for my second time ever on a surfboard.

Between the hostel and the local taco stand, it suddenly dawned on me that I was really alone. All the people I'd met travelling through the US were gone, left far behind. My time with Jamey in San Diego had been brief but intense and now I was back on my own. Not only that, but it would be this way for the next 6,400km, all the way to Panama. The language barrier meant even the small daily communication was about to dry up. So, while on one hand I was getting everything I craved and the adventure was getting more real, on the other, there was a sense of isolation to learn to deal with. It left me feeling a little anxious about what lay ahead and required effort to push that thought from my mind.

Back on the road, a physical obstacle forced me to sacrifice a couple of kilometres of running for a ride in a vehicle, the only time this happened on my whole adventure. The road between La Fonda and El Sauzal was closed due to a landslide. At the blockade I tried to persuade the guards to let me through and even became quite childish, refusing to turn around. During my futile protest, a pick-up arrived and one of the managers of the project gestured for me to get in.

Conceding to this alternative was pretty soul-destroying and made all the worse when the couple of kilometres we drove were absolutely runnable and not in any way dangerous. My mind was racing; would people judge me for doing this? This was probably the moment I came to terms with not caring what people thought. This whole adventure was never designed to break or set a record, or even get recognition from others, this was a conscious decision to change the direction and pace of my life. The terms were mine and mine alone. Deciding both to run Baja and to interrupt the adventure to fly home for my sister's wedding was only possible because I held this view and that is all that matters.

The reason for visiting El Sauzal was to meet up with Justin, Jon's surfer dude friend. He had a holiday cottage there, where he had enjoyed his misspent youth! What I didn't realise is that I would soon be experiencing a little of this with him. Justin arrived with a friend and a few neighbours gathered for drinks. Before long we decided to pop into Ensenada for some drinks and mischief. We jumped into the truck and beers were passed around. It turned out that our first mission was to find a pharmacy to buy some tramadol for Justin's friend, who enjoyed it for recreational use. Then we hit a few bars before ending up in a strip club called Paris de Noche. By this point, I was merely going with the flow, feeling a little uncomfortable, though undeniably enjoying the risk and danger element. Things seemed to get a little out of control and we were soon being asked to leave by a large, tattooed bouncer and, much to my relief, we were soon heading back to El Sauzal. However, when we got back things seemed equally surreal. New faces had appeared and everyone was descending into various states of booze- and drug-fuelled trances. While everyone indulged in whatever took their fancy, I took the opportunity to disappear to my room get some rest before the next day's running.

Ensenada was reportedly the last established town I would be passing through for quite some distance and I was advised to stock up on cash and essential supplies. While sitting in McDonald's calling loved ones and savouring the air conditioning, I was filled with a real sense that this was the true start of the adventure. The first few days in Mexico had been starkly different from the USA but the American influence was still very much present. Moving south, things would become more barren, the distances between towns greater and the need to be self-sufficient more pertinent.

My first day of running in the more desolate Baja California felt really exciting. The single-carriageway road wound through long expanses of nothing and the views were spectacular, with dramatic hills on either side of the sweeping valley floor. Passing through the small village of Santo Tomás, a man pulled over and just handed me an ice-cold Coca-Cola before driving off. This small act of kindness drove home just how much the temperature would factor into the journey from now on, and that finding shade and hydration would be crucial and at times difficult.

Over the next few days, I put my head down to clock up good mileage day after day. The best way to conquer this harsh environment was head-on, covering as much distance per day as possible. My thinking was that the further I ran each day, the quicker I could make it to places to resupply, rest and recover. Camping at the small farm shop, I meet the owner of a hotel in El Rosario, 180km south. He told me that the Baja 1,000 motorsport race would be passing through in a few days and when I got there I could stay at his hotel for free, which gave me purpose to keep moving.

Cash is king when you are in the middle of nowhere and that becomes no more apparent than when you can't take money out of the cash machine. This happened to me in San Vincente, a small village that happened to have an ATM. Not having notified my bank that I'd be in Mexico, my previous days' transactions had raised a flag and my account had been frozen. To compound this, I couldn't make international calls to my bank. Luckily, with the help of my family back home, we found a way to inform the bank the transactions were genuine and get my card reactivated. This seemingly insignificant mistake taught me to be more vigilant with my payment methods, something, in hindsight, I should have been on top of from the start.

Moving once again, the scenery filled my senses, as vineyards lined large areas on the side of the road. At the end of the long valley lay Punta Colonet. My hopes of finding a hotel were dashed, and once again it wasn't the kind of place where wild camping would be easy, due to how busy it was and the land being fenced off. I guess I also had the warnings of Mexico ringing in my mind! There was nothing for it but to try to find a hotel further down the road. Finally, in the punishing heat and with tired legs, I stumbled on a nice-looking hotel called Hotel Paraiso where the owner kindly gave me a discount. Other travellers had taken refuge there and, while eating cactus for the first time, I shared stories with an older Canadian couple who were cycling south.

Feeling refreshed, I was determined to make good ground the next day. The arid landscape with small twisters kicking up dust and tumbleweeds drifting over the road helped make everything feel even more adventurous. On both sides the land was flat, making distance harder to determine. The roads were also becoming straighter, adding a new, hypnotic dynamic to concentration, and I soon learnt to zone out and not let the surroundings affect my perception of progress. At first it was hard, but I set my sights on the huge stretches of road throughout the next 14,000km and the need to learn how to handle them. The trick is to take a new hardship and turn it into a lesson for the future. Now was no time to be defeated by smaller versions of the real challenges that were to come.

The distance ticked by, punctuated by interactions with locals. One day I bumped into a group of cyclists who had assembled for lunch by the side of the road. The hosts were the support team for a couple of impressively-bearded Mexicans who were cycling the length of Baja. Their father, who headed up the support, had set up a full kitchen and was serving delicious food, which they shared with me. It was nice to spend time with like-minded travellers who actually spoke English, and in such a surreal setting, even if just for a moment, I felt just a little less isolated.

On another occasion, my isolation was more problematic in the town of Vincente Guerrero, when a group of intimidating guys on motorbikes started hurling abuse. At least, that's what the tone implied. In these uncomfortable situations, you have no option but to keep your head down, not react and keep moving. It's essential not to show fear or weakness no matter how you feel. Luckily, in this instance, I had a place to run to and the hospitality I met there would quickly erase the memory of this encounter.

As the landscape became bleaker and less inspiring, moving forward became more of a challenge. The land offered very little shelter, and therefore almost no shade. At the roadside I'd stop and try to position my running stroller in such a way that it created some shade. But at midday, when the sun was at its hottest, even this was futile. At those times, I had to hope, even pray, that some shade would appear. These were some of the moments that I really had to dig deep and find that inner strength to continue—and this was starting to happen daily. The difficulty was finding ways not to focus on the negatives and rather, delight in the positives. On one such day, lunch at a fish taco stand in Lázaro Cárdenas afforded a much-needed break. It was a simple hut at the side of the road and the tables and chairs were plastic, but I stayed here as long as I could, continually going back for more. The thought of more running was overwhelming, but I was learning how my body, mind and motivation were reacting to the continuous flight mode I was in. Taking time to quiet the noise in my mind and refuel properly was working but I had to force myself to do it.

This resolve would be shaken over the next six days, however, as the landscape changed to huge fruit farms that lined the roads for 250km. Each night was a struggle to find places to camp and when I did, my energy levels would be barely enough to pitch my tent and cook a meal. If I could just notch up one more big-mileage day I'd be in El Rosario with somewhere comfortable to rest. The last 50km was made possible by people who pulled over to hand me cold drinks, workers who supplied high-vis clothing out of concern for my safety, and Les Misérables blaring on my stroller's speaker. As the hotel owner had described, motorbikes started to appear in the desert as I neared the Baja 1,000 race and more and more people could be seen lining the roads. It was here I experienced my first military checkpoint. Despite all the warnings north of the border, the soldiers weren't that interested in me as I was ushered through and on my way into town.

By now, the anticipation of seeing people again was lifting my spirits, especially as I thought there might be some English-speaking people around. Predictably, the hotel at which I'd been promised a room was full, but the staff let me camp in the car park, assuring me there would be a room the next night. These little setbacks are difficult to take after the exhaustion of running over 50km so letting them wash over me was the only way I could deal with them.

Despite being exhausted from seven days of running, I would let nothing prevent me from going into town and watching the race. The Baja 1,000 is a race from Ensenada to La Paz, essentially the route I was taking, but it was off-road and contested by cars, trucks, motorcycles and ATVs. By absolute fluke, I was in El Rosario on the exact evening the race was set to pass through. The streets were lined with teams of mechanics, supporters and locals. As the sun set, vehicles could be heard growling in the dusk before their headlights sprang over the horizon and raced across the main road. Every time one emerged, cheers would sound out before silence resumed in anticipation of the next competitor.

Evidence told me that my run would get more and more demanding the further south I went, and this thought was only accentuated when passing a sign on the way out of El Rosario that indicated that it was 318km until the next petrol station. Having spoken to people in town, I learned that access to things like food and water would become less reliable and further apart. To counter this, I changed my litre water bottles for gallon ones and reluctantly left any food behind that would melt or be affected by heat.

The town of Cataviña, some 130km down the road, became my next target. Knowing I only had three days of nothingness somehow made it seem manageable. Again, the road provided the entertainment, with a family pulling over to replenish my diminishing water supplies and an incident in which I impaled myself on a cactus while trying to pee adding amusement to my morning routine. At least it did in hindsight.

While ascending a hill I spotted a teddy bear that had presumably fallen off a car roof or trailer and had become entangled in a cactus. I scaled the barrier to prise him free and decided on the spot it was time to adopt a mascot for the rest of my adventure. Meet Carlos, named after my brother Charles. My new companion became my 'Wilson' and I started directing both my frustration and elation at him, rather than into nothingness. Although just a bear, he had relatable features and therefore converted my ramblings into a kind of conversation.

The middle of nowhere became the ideal setting for strange and wonderful meetings. At about 11am one day, a convoy of six young Alaskans pulled over ahead of me. As holidaymakers, they were as bemused to find me as I was relieved to see them. It didn't take long for a cool box to emerge and soon we were all sitting around enjoying beers and telling tales from our respective adventures. Their enthusiasm for what I was doing gave me more motivation and also the confidence that even amid the unbroken emptiness, there are always people. My fears of running out of water or food were allayed a little and this provided an injection of positivity.

Here in the desert, punctures had become a real issue and every time I ventured off the tarmac goatshead thorns would strike, especially at night when looking for camping spots. By now fixing punctures had become second nature, yet when you have to fix punctures in all three tyres, only to get more a few minutes later, especially at the end of a long day, emotions can get of the better of you. Uncharacteristic frustration coursed through my veins at these times.

It's true that for every difficulty, there is a silver lining and to compensate for my shattered nerves, the running just north of Cataviña was surreal and beautiful. The sky was a clear, deep powerful blue and on both sides of the road Cirio trees spiralled up between huge boulders. This otherworldly landscape used to be a seabed before it was seismically thrust up above the waterline. The effect was best described to me as if a giant had picked up a mountain, crumbled it in his hands and then strewn it across the land. It is a wonderful place and running through it felt like I was in some weird music video. To add to the effect, I listen to Queen's I am Going Slightly Mad as loud as possible.

Cataviña finally emerged from the desert as nothing more than a cluster of a few small buildings. Despite this, it raised a sigh of relief to have connectivity, convenience stores and restaurants after barren roads.

For my birthday, my little brother had given me some money on condition it was spent on a night in a hotel. On arrival in Cataviña I found a beautiful colonial-style hotel built around courtyards, one of which had a pool. What was I to do, but treat myself? The rooms were clean and better than I had seen so far in Mexico, so I still feel slightly bad that I promptly dismantled my stroller and used the sink and clean white towels to clean and mend the tyres that had been punctured a couple of days before. The number of punctures I had in Mexico was unbelievable and it felt like a day didn't go by without patching up more holes.

A lack of cash was weighing on my mind and at breakfast it occurred to me that I could possibly score some by approaching the other guests and offering to put their meals on my card in exchange for their cash. I set my sights on a couple of Americans who looked like they were part of the Baja 1,000. 'Is there any chance I can pay for your meal and take cash in return?'

'Sure you can, but why do you need to do that?'

I explained my predicament and they just stared at me. Then one reached inside his wallet, removed a crisp $100 bill and handed it to me. 'Look, you don't need to pay for our meal, just take the money. You deserve and need it more than we do!'

A couple on the next table asked what my plan was and when I informed them I had no option but to move on because I couldn't afford to stay another night, they proposed that I stay in a cabin nearby owned by one of their friends. Hell, yeah! Without hesitation I jumped at the opportunity to have a day off and a free night's accommodation, so we agreed to meet out front. They led me down a rough track to a small, one-room cabin in the middle of nowhere. It belonged to a man who allowed friends, and friends of friends, to stay in it on a first-come, first-served basis. My host-by-proxy also had an idea for how to protect my tyres from being punctured and he returned with a bottle of 'Slime'. This miracle is inserted into inner tubes and the tiny shreds of paper it contains catch and seal any thorn that enters the tyre so the air stays in. I spent the day relaxing in and around the cabin and enjoyed being away from the road. This tiny bungalow comprised a bed, chair and cooking area. Perfect! While I was there another chap turned up hoping to stay but when he saw me, graciously averred that I was there first and moved on. While the lodge was beautiful, I can't deny that the best thing was the toilet. Located about 200m from the hut, and apart from three ramshackle walls, it was completely open to the elements. It's hard to imagine my London self, pre-adventure, being quite so overjoyed at the prospect as I was that day and that was probably due to having to use some of the most godawful toilets imaginable.

Getting new kit supplies on the road was difficult as I never really knew where I would be or to whom items should be sent. Luckily, Baja did offer one solution and that was due to the constant trickle of Americans travelling south for the winter. I contacted Justin to ask him if he knew anyone making the journey and soon, we had a nice little strategy in place. I would get new kit sent to him in San Diego and he would offload it to friends heading south. The rest was in the hands of fortune and the assumption that they would have to pass me at some point. At least in the central part of Baja, there is really only one highway. So, as the sun set on the spectacular desert scenery one November evening, a convoy of cars pulled over to the side of the road. They were obviously in a rush because they quickly unpacked my rather large box and hastily jumped back into their cars, before disappearing south. Now stranded with an overloaded stroller, I precariously balanced the box on top for the last 10km to a small roadside restaurant. Here I unpacked my new treats from sponsors in the UK and started donating old kit to the workers who lived there. I had a rather amusing exchange with one man trying to persuade him why he might need a sleeping mat, the concept was so alien to him, but I soon convinced him.

The run to Punta Prieta ended up being one of the longer days of the adventure so far. With no intention of covering 62km that day, as I was approaching my intended campsite three cyclists passed me, a couple and then a solo cyclist. Explaining where they were hoping to camp, they suggested I join them if I could make the distance. Never shy of a challenge and fuelled by the prospect of the company of fellow travellers, I decided to make the extra effort. The last 20km were punishing, especially as the road was long, straight and with very little around but cacti, sand and sun. When I finally arrived in the village they'd mentioned, there was no way to contact the other travellers, so I had to find other options. One of the resident families ran a small restaurant and sold me some food, and I was about to set up camp in one of their front porches when I received an email and a location. Disappointment vanishing, that night I spent a really enjoyable time with Ryan, Peter and Allegra camping behind a restaurant where we ate and shared beers. There is a real camaraderie among the loners who seek adventure but also cherish the small opportunities when we all can come together to recharge spirits and inspire each other.

Further down the road in the hamlet of Santa Maria, I was permitted to sleep behind a small café, where the ladies that owned and ran it invited me to join in some birthday celebrations. It is these little moments of inclusion that affirm just how embracing and accepting life on the road can be. Little did they know that this little party coincidentally marked my 100th day on the road and just over 3,500km run.

Continuing south towards Guerrero Negro, the road started to get ominously straight, flat and featureless. The gaps between any form of civilisation were more protracted and shade was completely non-existent. After setting off in the morning you were exposed to the elements with scant places to rest. The heat rays shimmered as the road ahead stretched out into nothingness.

In Guerrero Negro, I managed to escape the heat for a day in a small hostel. Enveloped in home comforts, I ate well while attempting to recover for the next push. Ahead lay a stint of about 300km to Mulegé, crossing the Baja peninsula from west to east. My running shoes were starting to wear thin and I needed replacements or risk further injuries. Trying to buy running-quality shoes in this remote part of Baja wasn't easy as sports shops simply didn't exist. On this occasion, I had to settle for a pair that came wrapped in cling film and purchased at the local coffee shop, never really knowing if they were authentic or imitations.

CHAPTER 8 - PHONE LOST, FRIENDS FOUND

DISTANCE TO BUENOS AIRES: 13,424KM

The further I pushed into Baja, the more I realised that I needed to change my mindset. In America, it was possible to focus on where you wanted to end up on a day-by-day basis, but the more remote the road, the more you had to stretch those markers out and learn not to let the distance, time nor hardships affect you mentally.

As I left Guerrero Negro my plan was to run west, then along the coast of the Sea of Cortés to Mulegé, which I estimated would take me seven days. The journey would traverse from the long, flat roads of the Pacific Coast, across the ridge of Baja and down more undulating coastal roads on the east. The first day of this transition would follow a 100 percent straight road that pierced the barren landscape.

After 40km, I stumbled upon a small village and set up camp in a lorry park. Stray dogs sniffed around my tent all night, waking me from time to time. Imperfectly rested, I headed to El Marasal the next day, while a relentless wind forced the hot desert air into my face, encrusting sand around my eyes despite wearing sunglasses. The first relief came on the fourth day after a rather restless night camping in a deserted quarry. San Ignacio is a true oasis and its water, greenery and lushness seemed so alien to everything around it. Having only covered 18km, I was reluctant to stop but my senses got the better of me. Rather than camping on the side of the road, I decided to venture into the main village and explore the beautiful Misión San Ignacio Kadakaamán that stood watch over the village square. Spending half a day in such a beautiful and vibrantly alive village was exactly what I needed, and it was encouraging to find out that my presence here was becoming known to the locals. They had heard about my running on Facebook, or The Face as they referred to it, and were eager to photograph this strange visitor.

After a short day and a night in a proper bed, I felt sufficiently revived to tackle the 45km to Las Tres Vírgenes, named after a trio of large volcanoes that form part of the backbone of Baja. After leaving San Ignacio and re-joining the Pan-American Highway, I refound my enjoyment of running. At the top of the first hill I chatted to a Mexican cyclist who kindly gave me some excellent jerky. He was celebrating his 50th birthday with a cycle tour and we shared stories of our respective adventures before he sped into the distance, leaving me to refuel on strong coffee at a small roadside café.

Las Tres Vírgenes beckoned to me. All day they sat on the horizon and slowly grew as I got closer and closer. Aided by the gallon of water I'd been given in the middle of nowhere by some young men in a white van, I plugged on, only bolstered by the stunning views. Having spent what felt like a whole day running uphill, the last 5km were blissfully downhill. As the stroller's momentum dragged me onward, the Sea of Cortés ahead was a welcome relief. For the first time it wasn't the Pacific I was looking at and instead of keeping an ocean on my right, I now had to get used to an undulating sea on my left. Coming to a halt by a farm on the side of the road, I found a flat patch of grass to camp with a view of the sun setting behind the three volcanoes.

As the sky darkened a lorry arrived and revealed that the 'farm' was actually a metal recycling plant and its heavy machinery disrupted the tranquillity. When the work was complete, the men gathered around for drinks and traditional Mexican music replaced the clanking machines while the workers danced. I tried to imagine a scene like this in the UK but sadly have to admit it seems unlikely.

The run down to the coast the following day was perilous and steep. Not only was it hard work keeping the stroller from careering down the steep descents, but large lorries thundered past in both directions. A burnt-out car on the side of the road didn't add confidence and the coast, when I got there, turned out to be far more industrial than I had imagined, with the Boleo mine snuggled next to the shore. This is a copper-cobalt-zinc-manganese deposit that has been mined since 1860 and recently reopened as part of a South Korean project.

The local town was Santa Rosalía and I arrived there in time for lunch. As a longstanding mining town, there is a real Wild West feel to it. Wooden buildings line its streets, monuments to its past, with a prefabricated church said to be designed by Alexandre-Gustave Eiffel the star attraction. The Iglesia de Santa Bárbara was first shown at the Exposition Universelle in Paris in 1889, then moved to Brussels before being bought by a mining company and shipped to Baja. A funeral was in progress so I couldn't venture in but even the outside is captivating, with the bottom half made of wood and very traditional, while the intricately gabled roof is white, industrial and metal. Even now, it has the air of an exotic visitor to Mexico.

With no cheap accommodation in town, I ran on and received a surprisingly frosty welcome in Ejido San Lucas from the owner of the RV park. He operated strict conditions about who was allowed to stay, preferring 'snowbirds'. These are people who migrate every winter to escape the cold in the northern US and Canada and head either to warmer states such as Florida, California and Texas or leave the country altogether and hit Mexico or the Caribbean. Fortunately, Rene, the local restaurant owner, and his wife very kindly allowed me to camp on their plot and I settled into a night listening to guests performing live music while I ate amazing fried chicken cooked by my hosts.

Mulegé was just 48km away. The prospect of even one day's rest there provided all the oomph I needed. As another oasis town, it has a reputation for being a place that people enjoy so much they have trouble leaving. Mulegé has long been a popular destination with Americans, helped by Hollywood stars such as John Wayne and Olivia Newton-John visiting the Hotel Serenidad. Nowadays it seems to be a place where Americans can live a lifestyle they may not be able to afford north of the border. I found a little cheap hostel with basic rooms and, not having had a shower for three days, was overjoyed to clean up and head out for a proper dinner, even amid a town-wide power cut.

The harshness of the Baja desert and the weight of the additional water I was pushing was taking its toll on my stroller so I set about finding mechanics to help get everything back in order. The first issue was that the front wheel was not spinning as it should. On investigation, all the ball bearings had fallen out and the axle was grinding with every turn. In the local hardware shop, a man called Chris intervened to say that he knew someone who might know how to fix it. After a failed attempt at a cycle shop, we were referred to the guy who swept the town plaza and lived out of town. Chris very kindly drove me down the hurricane-mauled roads while giving me the lowdown on Mulegé. As we chatted, a majestic osprey hunted fish from the Rio Mulegé and while soaring back into the sky, adeptly repositioned the fish in its talons to be more aerodynamic before heading inland.

When we finally arrived, Antolio took one look at the wheel, nodded, dropped everything and set to work. Antolio was a short man with a kind face marked with a large scar above his left eye. I don't know the full story but apparently, he had sustained a severe injury and thereafter led a solitary life helping out around the town. He appeared to know what he was doing and very quickly my wheel was spinning freely once again!

Despite the temptation of staying put in Mulegé, in part due to the local bakeries, I diligently set about packing for the next day. That was when disaster struck, as I discovered my phone had disappeared! It had been in my pocket but at some point that morning had fallen out. That phone was everything to me—mobile phone, camera, social media tool, blog updater and source of music. Our lives orbit our phones anyway, but thousands of miles from home and without any other means of support, this seemed an even bigger deal. Without delay I set off in search of it, retracing every step back to Antolio's and the other mechanics. It was nowhere. I even tried the 'Find my phone' on my laptop but due to a weak signal, had no luck. As a last resort, I visited Chris at his home in case it had fallen into his buggy, but alas no. There was no option but to stay in Mulegé another night. It certainly seemed difficult to leave this town, although not for the reasons I had been warned about.

It seemed like a major setback at the time, but the disaster translated into some great experiences. First, I bumped into Jessie. We'd met in the Redwood Forest where he had talked about cycling south and boom, there he was. Still reeling in amazement, we made plans to meet up in a couple of days if our journeys crossed paths again.

I retraced my steps again and again but to no avail. Stopping at a bar to check my email, 'Find my phone' had miraculously done just that! In complete ecstasy, I jumped up and asked if anyone knew where I could get a taxi. That's when I met the amazing Tom. He immediately offered to drive me to try to recover it. His worry was that the house in question was where the local bad boy lived. Disappointingly, when we got there no one was home. On the way back we stopped at a bar to devise a plan to get my phone back. We met the owner, Carlos, and soon we were all standing at the bar with Scottish pipe music blasting out—the barman for some reason loved Scottish music and was delighted to have a real Scot in his bar.

After this impromptu night of drinking and general silliness, we arranged to drive out to the house again first thing the next morning, so at 7.30am I met Tom for breakfast, as well as a friend of his called Steve whose amazing stories included times with John Wayne, helicopter adventures during the Vietnam war and the charity work he was doing in Baja.

On our way to collect my phone, we met the wife of the guy who had it. She said he was at work and would bring the phone to me when he returned. My only option was to wait for him to contact me. That afternoon he duly drove into town and handed the phone over, for which I rewarded him with $20. He told me he'd found it lying on the road that led to his house; being bright yellow it was hard to miss! Drama over, relief flooded over me and I was all set to leave Mulegé. However, my joy at being back on the road didn't last long, as just a few hundred metres out of town my front tyre exploded. I changed the inner tube and continued on my way only to be halted a few hundred metres further down the road. On closer inspection it was obvious that the tyre rubber had worn completely through, exposing the inner tube to the asphalt. This was serious, as without an inflated tyre, how could I push the stroller at any speed?

While figuring out what to do a large American truck screeched past and a woman shouted out of the passenger window, 'What the fuck are you doing, you crazy idiot?' Slightly startled at first, I quickly realise that she must have thought that there was a real kid in the stroller. Perhaps many other motorists along the way had come to this same conclusion. This amused me enough to briefly take my mind off the problem at hand.

Still seeking a solution, I visited an American I'd met over dinner in Mulegé. Maury lived in a small beach community called Concepción and had been in town stocking up on supplies. After a quick tour of his place and a cold drink, I took the road south to a beach called Los Cocos, hoping to meet up with Jessie. As I left, Maury handed me two small water bottles and a plan for how to fix my tyre formed as I struggled forward.

Limping my stroller on to the horseshoe beach of Los Cocos, Jessie welcomed me to a thatched shelter made from palm leaves called a palapa that would become my home for the next couple of nights. He then introduced his new friends Deano and Jackie, who were based out of a camping car called the Vaca Loca, complete with a horned cow skull decorated with shiny glass beads.

Conscious that I would not be able to relax until my stroller was fixed, I set about tending to my tyre issue. Working with Deano, we cut the small water bottles that Maury had given me and then covered them with black electrical tape to cover any sharp edges. The curve of the bottles fitted perfectly inside the tyres, acting as a barrier between the road and the inner tube and the pressure created when the tube was inflated meant that the patch would remain in place. The next town was 110km further on and there I hoped to be able to purchase new tyres and return to full strength.

That night, Jessie, Deano, Jackie and I attended a drinks party hosted by other snowbirds. This wasn't a term I'd come across before Baja but I was sure meeting lots of them now! Guests included a couple, Paula and John, who had decided to pack up their comfortable lives in the city and exchange it for life in a campervan.

Mulegé hadn't afforded me much time to relax, however, in Los Cocos that seemed to be all anyone did. Jackie and Deano took Jessie and me under their wings without hesitation. As the sun rose, Deano wandered over with freshly brewed coffee, sweetened with delicious honey. Later that morning, a local fisherman sold us fresh fish, which Jackie and Deano made into ceviche to share, accompanied by cold beers. The afternoon was spent snorkelling with their kit and spending time with Jessie as he readied to depart.

Deano was one of those people who could forever regale you with his stories. As he sat there in his board shorts and tee-shirt, sipping a beer through his long moustache he would recount the time he could have been on the American high board diving team but got banned for competing for money. No matter the subject, you couldn't help but be drawn in. Did I question some of them? Of course, but I wanted to believe that he was as colourful as he was kind and generous. Unusually, I didn't succumb to that strong urge to get back onto the road so extended my time there by another half-day. Next afternoon, after more ceviche, I reluctantly said my goodbyes and returned to the solitude of the highway.

The road south to Loreto was beautiful and undulating and I felt a little homesick as it hugged a bay that curiously reminded me of Loch Lomond back in Scotland, even more so when I started playing Highland Cathedral on my small speaker. It's strange, but when completely alone and isolated, loneliness seems to bite less, but after sharing good times with people, the mind finds ways to bring those feelings to the surface.

Deano and Jackie were also making their way south and even managed to bring a little culinary bliss to my day by taking lunch orders and then stopping 10km down the road and having everything prepared as I arrived. Little interactions like this meant so much to me but I never felt I could convey my gratitude. Deano and Jackie were becoming a regular presence, an interesting dynamic as normally, new connections swiftly just became fond memories. It was refreshing to be building a bit more of a relationship, even if it too would have to end.

As much as I had enjoyed the comforts of the last few days, returning to the road and to camping summoned me. It was mostly featureless, although one evening I managed to find a small restaurant and negotiated permission to hang my hammock between the pillars of the building. Eventually a Mexican cyclist called Hugo who was also heading south joined me, and we shared the shelter. Astonishingly, the restaurant didn't serve food but had an abundance of beer. A night of story sharing and beer, with very little actual nutrition, ensued, enlivened by a van full of very drunk locals trying to buy beers in the middle of the night.

The road to Loreto seemed endless with the sun relentlessly beating down on its many hills. Despite getting closer, my suffering mounted and it even entered my mind that I might not make it to my destination. Immediately identifying the risk this negativity posed to my performance, I somehow dug deep and found some positivity. On finally stumbling into Loreto, I was running on empty. I looked up restaurant recommendations and proceeded to eat myself back to normality at a roadside restaurant serving massive American portions.

Despite the mental and physical toll on me, I was very aware that my stroller was running in a sub-optimal state and that had to be the priority. Without it, there was no adventure and as much as I was focused on keeping myself in the best possible condition, more time, energy and investment needed to be spent on stroller maintenance. In the heat of the day, I trudged around the dry, sandy streets of Loreto, trying to find a decent bike mechanic and check it in for a full MOT.

My couple of evenings in Loreto were spent in a small RV park where a Los Cocos reunion was gathering, plus a whole host of new faces, all seeking some respite from the harsh desert road. Jackie and Deano were my hosts again, but Paula and John were there too. We were also joined by Emma and Bill who had a big four-by-four and called themselves the Flightless Kiwis, and a Latvian couple with a converted van.

With my stroller in the workshop, another day's rest was imposed on me and spent refuelling for a final push to La Paz. There I would have to address the challenge of actually crossing the Sea of Cortés. With just 360km left, I made the decision to do it in one push. Unusually, four of the last 16 days had been rest days, and while not all of them were restful or planned, I worried I was underachieving. At La Paz I knew I would be forced to stop for a few days, though it was unclear at that moment just how many. Unfortunately, the final stretch wasn't a simple dash down the coast but rather a large curve over to the west, crossing the Gigantes range. After a final night camping with Deano and Jackie on a small beach near Ligüí, I prepared for the long climbs to long stretches of nothingness.

Hills posed a few challenges beyond the obvious ascent. In these remote desert landscapes, the amount of food and water required increased the weight of the stroller to about 40kg, which was hard to push uphill. Then on the descent, that weight pulled the stroller down and with no brakes, holding the stroller back was hard work. I will admit that if I were to pack now with the knowledge I have picked up over the years, things would be different but in the moment, I had no option but to grit it out and literally roll with the challenges.

In the town of Constitucion, breakfast was had at a roadside taco stand. A local gentleman was intrigued by my adventure and it didn't take long for him to start sharing his own adventure stories. A quick rustle around in his glove box produced some photos of his cycle tour of Africa in the 1970s. We shared stories as we gorged ourselves, filling our bellies and our imaginations. As crazy as it sounds, listening to him talk about his travels in a different era made me realise just how different technology has made adventuring today. I envied the purity of his journeys. No ATMs, smartphones, internet or GPS, just a bike, a dream and a destination.

For the next couple of days the road was as straight as a Roman road, with no hint of a curve or deviation. I met another cyclist, this time from Japan. His name was Rio and he was partway through his dream of cycling in every country in the world. While he'd already visited over a hundred countries, he admitted that his Japanese passport was going to make it difficult to achieve, but that did nothing to deter him. His policy was to keep going anyway, bag the countries he could and tackle the harder ones later down the line. I loved that a future obstacle didn't dissuade him from his goal and it reassured me of my own approach.

Gradually but noticeably, the land was changing again, becoming hillier and with more scrubland on either side of the road. I knew I was not going to make it to any civilisation by the end of the day, so it was a case of finding somewhere to wild camp before sunset. While scouring the verges for somewhere suitable, I had a little spiritual encounter thanks to a priest who was walking his dog on the side of the road. As I passed him he produced some fruit, a cold drink and a wooden cross pendant. He then blessed me before reaching into the cabin of his truck and pulling out a can of beer, which he memorably drank in one, before jumping into the driving seat and accelerating away.

This part of Baja was known for being a bit dangerous and I received constant warnings about banditos. Being on my own did make me feel vulnerable on occasion and today was one of them. Trying to find somewhere discreet and sheltered to camp was proving difficult and when I finally found a little used track, I set up camp. In the fading light I became aware of voices in the vicinity. Like a scared animal, I hunkered down in my sleeping bag and remained as quiet as possible to avoid giving away any clues to my existence. It's strange, but in the twilight, the very presence of strangers evokes a sense of danger, and worst-case scenarios jump to mind. In the light of a new day those same voices were as close as the night before, but the threat seemed to abate with the realisation they were just doing the same as I was.

While finally in reach of La Paz, the knowledge I would finish at sea level made the remaining 55km seem more palatable. The lure of a hostel, interaction with people and a break with beer were all positives. 350km in seven days was ambitious however I looked at it, but the realisation that I could pull it off was enticing. The final stretch, as always, was not as straightforward as I had hoped, with long stretches of abandoned roadworks forcing me onto sand tracks that took far greater effort than asphalt. Military checkpoints and properties guarded by ferocious dogs that chased me down the road also added to the day's excitement.

CHAPTER 9 - HITCHHIKING OVER THE SEA

DISTANCE TO BUENOS AIRES: 13,424KM

La Paz turned out to be a lot bigger than I had imagined and jogging through the suburbs felt like forever, punctuated with the occasional stop for ice cream. To celebrate finally arriving at the harbour was the prize of a hamburger and a nice cold beer. Between mouthfuls, I thought about what I'd achieved by running nearly 4,400km in 125 days. It felt impressive but, possibly because being on my own was becoming the norm, I didn't feel like I could share it with anyone. For some reason, I didn't think that anyone would care. Looking back, perhaps I should have made more of these milestones because, as I would later find out, people really did care!

La Paz was a shock to the system. Having been pretty much alone for the last month-and-a-half, with very little sustained interaction with people, here was a large vibrant town with every possible convenience and people to meet. Clearly, running was not going to get me across the Sea of Cortés, the body of water that lay between me and mainland Mexico. But before dealing with that hurdle I decided to treat myself to some diving, an old hobby of mine. Having gazed at the water for so long, it was time to take the adventure underwater.

At a diving centre I planned an eye-wateringly expensive dive for the next day. When you are living as cheaply as possible, big-ticket items suddenly become more imposing, especially when not necessities. However, diving has always been a passion and the chance to see hammerheads and whale sharks made the decision quite easy, if still a little painful.

My companions on the dive would be a motorcycling American Mormon from my hostel and two other experienced divers. As with every diving adventure, the journey to the dive spot was thrilling, with the sea spraying our faces as we bumped along, full of anticipation as nature-spotting sports involve a degree of luck and therefore disappointment. Once kitted up and descending, we were rewarded by sighting a small shoal of hammerheads far below, who promptly decided they were not in a sociable mood. Our second dive wasn't so lucky but the thrill of the first glimpse was enough, especially in the knowledge we were almost definitely going to see whale sharks later that day. The real highlight was our way back when our captain glimpsed something out of the corner of his eye and quickly diverted the vessel. We scanned the horizon and there it was, a huge humpback whale breaching the surface. I can't put into words the honour and privilege of spending time so much closer to whales than I'd ever achieved along the United States coast. The words majestic, graceful, beautiful and elegant just don't do it justice. These moments alone were worth every penny.

To cap it all, the whale sharks were where expected and three of their number floated mere metres from the boat as we tumbled over the side in our snorkelling gear. There was just no preparing for just how big they were and how they made me feel. Being in proximity to something so vast and powerful yet relaxed filled me with a sense of perspective. Because they were feeding, they were vertical in the water to suck in the nutrients at the surface. If things weren't perfect enough, a shoal of devil rays came gliding past to crown an amazing day out.

It was time to refocus on the problem at hand, namely crossing to Mexico. From what I could glean from the internet and other travellers there were three options. First was flying direct to Mazatlán but as it was Christmas, flights were expensive and availability low. Jumping on a plane didn't sit well with me so I immediately disregarded this option. Second was the ferry to

Mazatlán, but again availability was sparse until after New Year, meaning lots of wasted time and money in La Paz.

The final option was the most appealing. Some travellers had mentioned that you could ask at the marina if any of the sailors were planning on making the crossing and if they would be open to giving a lift, essentially hitchhiking on boats! The elements of uncertainty and adventure appealed to me and I made it my business to learn more about the daily morning meeting. It wasn't in-person but over the radio, with a handset for those on land to add requests during the 'any other business' item on the agenda. If I was expecting this to be formal then I was mistaken as the questions were more along the lines of immediate needs, such as where to buy horseradish sauce to make a proper Bloody Mary over Christmas.

I was warned that it could take a few days or even weeks to get a bite, but I was determined that this was how I was going to cross the Sea of Cortés.

My pitch the following morning at the marina was once more greeted with nothing more than a crackle of static on the radio. As the crowd dispersed a large, formidable-looking gentleman approached me. 'Are you the runner looking for a lift?' he asked in an American accent. On confirming that I was, he asked me to follow him. On stilts in the boatyard was a 50ft converted fishing vessel with the word Andante across the helm. 'This is my boat. She's just finished having some work done and we'll be setting off in a few days. You are more than welcome to join, though you need to meet my son, Zac, and he has to agree.'

To say I was bowled over by this was an understatement, so I asked, 'Why are you doing this?'

'When I heard you were raising money for a male suicide charity, I wanted to help.' It turned out Kevin had had a brush with suicide, and we would talk more about that later.

The next day at Kevin's boat, now moored in the marina, I met Zac. On obtaining his approval, we hashed out the finer details and agreed to start our crossing on the 23rd December, meaning we would spend Christmas day together and at sea.

Over the next couple of days, I relaxed and checked over my stroller and kit. My father wired me some money over for a pre-embarkation big Christmas dinner. Having made a few friends in the hostel, I used the money on a group meal at a local burger bar. To be honest, it didn't create the comradeship that I expected and unfortunately most just seemed happy to eat a free burger then disappear. Luckily, Kat and William, a couple of British travellers, got into the swing of things and we ventured to the local cocktail bar for festive drinks.

With the packing done the next day, and everything loaded onto the Andante, all that remained before departure was a huge shopping excursion to supply us across the Sea of Cortés. Kevin's generosity went to another level when he wouldn't let me contribute and even took us out for a last land-based meal.

Our plan was to moor in a small bay just out of La Paz to bond as a team, relax and teach me some basics about life onboard. From there we'd head directly to Mazatlán where I would disembark and continue my run. If everything went to plan, I would be running again in three to four days. Perhaps the boat's name was portentous, because it would in fact be 10 days before I was running again after a little unplanned maritime diversion.

Two days before Christmas, we left as planned. The weather was great and we moored in the bay to do some snorkelling. Later we bonded over Kevin's amazing margaritas, which he served in old pickled-onion jars.

Kevin was a big-hearted man who had achieved a lot in his life. At some point, he had run a coffee chain in Seattle and I got the impression it had been quite successful. He'd also captained a firefighting unit in Seattle and been active during the Twin Towers disaster. Responsibility for removing the dead bodies of fellow firemen from the rubble had fallen to him, leaving him with PTSD. He reached a very dark place, wanting to take his life as a result, so sought help and some time away from Seattle and work. That is why he was in Mexico and why he found my charities so interesting.

His son, Zac, was a huge, but amiable chap who was great around the boat. His relationship with his father appeared fractious at times, and part of the reason he was there seemed to be for them to bond.

That evening another boat moored nearby and we used the tender to motor over and introduce ourselves to the couple and their daughters. An impromptu drinks party soon started! The sailing community seem to love to entertain and although I've been around boats my whole life, I'd never really done anything like this, so it was a great introduction.

Christmas Eve began with a hangover combined with still getting my sea legs. Both Kevin and Zac were true boat people, and it was evident I was not, so I was glad neither saw me being sick over the side of the boat. It meant a lot that I wouldn't be seen as dead weight. This was made difficult when fishing off the side of the boat, as the only thing I managed to catch was myself. The barb of the hook went deep into my forearm and I had no choice but to reluctantly accept help by way of some wire cutters and iodine.

Being on a small boat in close proximity to two other people was a whole new world. The last 130 days had been spent alone, answerable to no one and responsible only for myself. Now I found myself in confined quarters with two strangers, doing something I wasn't trained for. It must also have been strange for their father-son dynamic, so I took the only opportunity to head to the beach to give everyone some space.

Evening came and we headed out of the bay into the Sea of Cortés towards Mazatlán. Once clear of the bay, the boat started to pitch and roll in the waves as the weather grew angrier. My role on board provided an extra person on shift, specifically from midnight to 4am, which was not far away. The further we got, the more aggressive the sea became and after a quick team meeting we sensibly decided to return to the shelter of an island and wait for the weather to calm down. One of Kevin's great qualities was not making anyone do anything they weren't happy to do—safety first all the way.

By 5am on Christmas morning, the weather had calmed and we readied the boat for a second attempt. The mood was very relaxed, with nothing especially Christmassy on board and strictly no drinking during the crossing. The whole episode had a rather unreal quality.

Boat life was very calm, and our routine ensured we progressed as safely and quickly as possible. During the day we cast a couple of fishing lines in the hope of catching dinner and Kevin and Zac rigged an empty beer can to the line to alert us if we snagged anything, but it only produced a couple of false alarms. During the day we were all on duty, cleaning and working wherever and whenever possible. In the evening, we took shifts to watch the radar and try not to collide with other vessels. Kevin took the first watch, I was second and Zac took the third. Everything went very smoothly and our only visitors were the dolphins that danced in front of the bow and the birds that hitched lifts on the rigging. Occasionally a butterfly fluttered by, which still blows my mind.

On the final morning of our crossing, a whale escorted us as we sailed. At the port we immediately set about the essential tasks, such as cleaning the salt off the boat to guard against rust. Kevin was particularly good at getting us to work and Zac and I enjoyed working together. Tasks complete, we went for a stroll around the town that afternoon, with everyone splitting off to do their own shopping. The truth is we all just wanted some time alone after a few days on the boat. Being back on land did the trick, as the beauty of Mazatlán's huge market and lovely colonial buildings ignited my passion for Mexico and the journey ahead.

After regrouping, we met a friend of Kevin's and enjoyed a few drinks back at the boat. During dinner, Kevin and Zac talked about their onward plans and suddenly I sensed that my time on the Andante wasn't over. They invited me to go with them to Puerto Vallarta via Isla Isabel, or the 'Galapagos of Mexico', saying that they could drop me in the small beach town of Chacala, a little further down the coast. I wrestled with this, asking myself: would it be cheating? Should I run? The only answers I conjure were excuses, such as flying would have taken me even further south to Puerto Vallarta anyway, or that as I'm not attempting to break any records shouldn't I take advantage of this opportunity? At heart, I was looking for ways to justify joining them on the next stage of their adventure, so there was nothing for it but to raise our glasses to the next stage of our journey together.

The plan was to spend one more day exploring Mazatlán and its beach before motoring overnight towards Isla Isabel. As the sun turned red and started to sink to the horizon, we bought margaritas then explored the islands by inflatable boat, before returning to the Andante. That night I was on the midnight shift on the way to Isla Isabel. This time the sea was calmer, but it tested my skills of navigation when another vessel was clearly heading straight for us. At sea things happen very slowly, and you make small adjustments in the hope of passing with sufficient distance between boats. As soon as you think you've solved the issue, the current changes and you're back on a collision course. This went on for an age until Zac came to relieve me and it became his problem, not mine!

Early in the morning we set anchor at Isla Isabel. Even from the sea, the island's beautiful crescent bay and golden sands were everything I had hoped for. A couple of other boats were moored close by and while the island is officially uninhabited, there was a small, illegal fishing settlement.

This was the closest I was ever going to get to a Jurassic Park experience. Our tender beached on a small patch of sand and we pulled the boat up onto the shore. The three of us headed into the tree line and stumbled upon the abandoned research centre. Although the building was derelict the illegal fisherman were evidently using parts of it as living quarters. Nonetheless, we roamed freely around the buildings and found iguanas lounging everywhere or scuttling around the site.

Ascending the hill behind the building, we spotted large numbers of blue-footed boobies and frigatebirds. Boobies' feet are bright turquoise and so unusual they don't seem real. I'd seen frigatebirds in the Galapagos Islands back in 2008 but here they were on the ground, in the trees and circling overhead. Words cannot express my elation just walking around and gazing at the majestic birds. A real treat, and if I had needed any justification to stay on with Kevin and Zac when first invited, now I knew I'd made the right decision. In awe and happy, we returned to the boat for snorkelling, fishing and socialising with the other boats' inhabitants.

Our next stop would be Chacala, where we decided to spend New Year's Eve before I set off south on foot. On our way to the mainland, we finally caught a fish after days of trying. Having fresh fish to eat was super exciting, so we immediately set about gutting and filleting it. Life onboard the Andante was good.

During the trip, my conversations with both Zac and Kevin were really constructive. Out of respect to them I want to keep them confidential, but it was an honour to get an insight into these two remarkable men's minds and share our thoughts on some heavy topics. Without realising it, we all had a lot in common and although what they were working through seemed to exceed how I felt, I could empathise with them both and also learn from them. Listening to them forced me to take a look at who I was, what needed to change and how this adventure could be a journey of discovery. This was the beginning of a deeper self-evaluation that would take place between this point and Panama.

At Chacala, just north of Puerto Vallarta, we moored the boat straight off the beach. Chacala's tourists are mostly Mexican, not gringo, which creates a nice, relaxed vibe. We made the most of it by celebrating the end of our little sea adventure together over dinner in the village. The next morning, slightly hungover, we drifted back and forth from the beach, blissfully not needing to be anywhere, or do anything. Except welcome the New Year, that is. For New Year's Eve we headed to the bars alongside the large Mexican families who had congregated there, parents and children smashing piñatas open with that mixture of exuberance and violence that the situation demands.

The main festivities happened on the beach where absolutely no health and safety precautions were taken. Everyone brought fireworks and it was just a big free-for-all. Some fireworks launched as planned and exploded, some fizzled out and some just hurtled along the beach. I stood watching in disbelief and wonderment. Everyone was so happy and drunk, with fire at their fingertips. The finale was a huge structure emblazoned with Catherine wheels and pyrotechnics that hissed and fizzed to the delight of the crowd. It may all have been very low-budget, but the atmosphere was verging on magical. For me, this was the perfect New Year's Eve.

CHAPTER 10 - ¡VIVA MEXICO!

DISTANCE TO BUENOS AIRES: 12,627KM

By the first day of 2015 the atmosphere on the Andante was becoming a little strained. Our New Year's Eve celebration may have been one party too many and we all woke with bleary heads. Zac immediately dived overboard with a dry bag and disappeared. Today was clearly a good time to get back on the road and leave Kevin and Zac to their bonding. I felt so indebted to Kevin for his generosity and hospitality so didn't want to overstay my welcome. With my stroller packed, Kevin gave me a lift to shore while we made plans to meet in Puerto Vallarta in a few days. Wanting to say goodbye, I found Zac sipping coffee in a café, by which time he was in better spirits. No doubt the combination of three men and alcohol in a confined space had started to take its toll on us all, but I was glad I got to see him before I left and to thank him for everything.

Getting back into running was made more challenging with a hangover and the fact that I hadn't run for nearly two weeks. During the first three months of my adventure I'd come to appreciate my own space, so the week confined on a boat made that time feel more precious. Although mainland Mexico wasn't technically a new country, this marked the start of a new section of my adventure that would take me 4,650km to Panama City.

A new deadline also loomed. My sister and her husband-to-be had finally set the date of their wedding and it was 27th June 2015, which meant needing to get to Panama City to fly back on 19th

June. At that time, I remember feeling that 4,650km in 170 days seemed like a lot to cover, however thrilling the challenge.

From the desert landscape and dry heat of Baja California, now I had to contend with dense forests lining the undulating coastal road. In the high humidity, I sweated profusely. The mainland somehow seemed edgier. On Baja, there was a general sense of relaxation but here everything was more frenetic.

After 30km, I arrived in La Peñita de Jaltemba to find another town that was geared towards Mexican tourists. Cheap accommodation could be had in a hotel that was full to the brim with Mexican families. Its carefree vibe may not have been that restful, but it was infectious as loud music played and children ran amok.

Kevin had recommended a night in Sayulita, so I heeded his advice even though it was a little out of my way. If this had been a holiday, the village's blend of hippy traveller vibe, local Mexican fiesta and good amenities would have enticed me to stay longer. Music filled the air as a group of singers busked on a street corner to an appreciative crowd. Everything seemed so chilled. However, after 12 days off and only two days under my belt, I simply had to press on.

Antolio's repairs had started to play up and at Mezcales, my front wheel seized up altogether. As luck would have it, I found myself right next to a small bike shop run by a friendly and rather attractive lady. While grappling with the cover to get into the ball bearings it occurred to me that I hadn't spent enough time learning the art of stroller maintenance before I left the UK. About three hours later we managed to disassemble the wheel and replace the ball bearings, before I called it a day.

The next day at Nuevo Vallarta there was no way to contact Kevin and Zac without contact details or the code to access the piers at the marina. It was sad and frustrating but I had to accept I would probably never see them again.

On approaching Puerto Vallarta, my phone beeped. It was my father, who was following my progress step-by-step. On this occasion, he was alerting me to the proximity of a Starbucks. But on arrival, I was shocked at how expensive things were and decided to restrict my spending to necessities.

Puerto Vallarta is a beautiful and colourful town and I soaked up the atmosphere of its bohemian old town for a few hours. By evening, the restaurants were buzzing and couples wandered, hand in hand. While these moments were enjoyable, they reminded me that I was alone and a misfit. The more developed and affluent the area, the more out of place I felt.

The next morning my mood darkened after a long and punishing climb to the outskirts of town where the last ATM had no cash. My huge verbal outpouring of frustration must have caused serious amusement to the passers-by! Luckily the running calmed me down as I navigated thin roads hugging the cliffs past the luxury hotels and large private houses that grandly surveyed the Pacific. While I had enjoyed Puerto Vallarta, I wanted something less commercial, so would cover only 15km that day on my way to visit Boca de Tomatlan. Boca, as it's known by the locals, is a small, quaint fishing village, where a handful of restaurants cluster around the shore.

Satisfied by a delicious fish and prawn ceviche lunch, I set about finding somewhere to sleep that night. One of the restaurant owners agreed that in return for eating and drinking there that evening, I could keep my stroller in his kitchen and hang my hammock in the restaurant when it closed. Looping my hammock between two beams of the restaurant's outside seating area, I prepared to bunk down for the night when a local chap who'd been hanging out on the beach during the day came over for a chat. Midway through telling me about the time he'd spent in the USA, his life as a gang member and a stint in a US jail, he suddenly sprinted off. A sickening thud from the shadows ensued, followed by an animal's cry. Minutes later he reappeared holding an opossum by the tail, which he promptly threw onto the ground before continuing his story as if nothing had happened. To add to my horror, the animal started twitching and was clearly not dead. The man casually picked it up and proceeded to beat the poor opossum's head against a rock, spraying blood everywhere, including on my hammock. When I asked why he had done this as dispassionately as possible, especially after what we had been talking about, he very calmly replied, 'Breakfast!"

On Baja, the openness of the landscape had allowed me to stretch my daily distances, which made me feel like I was accomplishing more. Here on the mainland, the private land, villages and geography made things feel more disjointed. I needed to kickstart my routine again to leave the traveller frame of mind behind and rediscover the adventure athlete. Here, even covering big distances didn't feel productive.

One night at a small local festival in El Tuito, I resolved to increase my daily distances and target the small village of La Cumbre, some 55km south. Predictably, several distractions would interfere along the way with this re-acclimatisation.

The first was the disturbing discovery of a shopping bag full of dead kittens. Death haunts the Mexican roadside, but this despicable act reviled me as it was clearly intentional. Later that day, I spotted my first crocodile from a bridge above the Rio Tomatlan. Two Mexican fishermen casually went about their business mere metres from this huge predator, swinging a circular net around their heads before launching it into the river. The net opened into a spiralling circle as it fell into the water, before being reclaimed and checked for fish.

While refilling my water further down the road, I met the owner of a vacant property in La Cumbre, who kindly gave me permission to camp on his land. Once there, I was showered with amazing hospitality from the couple who were the property caretakers.

Running from La Cumbre to Perula was challenging, as the heat and humidity felt like a wall around me. That, along with the many small ups and downs in the road, conspired to make the going very tough, both mentally and physically. I hobbled into Perula after spending the last 45km of the day with pain spreading through my right knee. Resorting to ibuprofen to numb the ache and with a potentially serious injury developing, I had no choice but to take a break. Admittedly, there are worse places to rest than the beach town of Perula, but I couldn't switch off. Even though I was injured and the right thing was to rest, an inner voice was telling me to get going, to cover miles and to stop spending money.

After two days I forced myself back on to the road and it was immediately clear that my knee still wasn't at full strength. Stupidly or naively, I persisted for 35km until the village of Emiliano Zapata, where I had to admit to myself that I couldn't go on. Finding a hotel, I asked Facebook friends for help and was rewarded! A physio friend-of-a-friend used photos to recommend some exercises to relieve the pain. Injury was always going to be one of the biggest risks in this adventure and the persistent problems in my knee made me wonder whether I could overcome this one. It even dawned on me that this could potentially end the whole adventure, but I refused to give in to self-doubt and did my best to shut that negative thought down.

The following day, I resorted to walking to protect my knee. The next town of any decent size was Melaque, some 51km away, and while pressing on may not have seemed the most sensible recovery plan, at least I felt productive. Having had two recent sombre encounters, little did I want another so soon, but on leaving the village I came across my first truly heart-wrenching experience. A flock of vultures were in a field, jumping all over an animal. On investigation I realised that the animal was a half-eaten calf and it was somehow still alive. Panicking, I immediately scared the vultures away then frantically searched for the farmer, or anyone who might help, but found no one. Every time I ran back to the poor animal to scare the returning vultures away, it looked in a worse condition. I wrestled with whether to jump in to put it out of its misery. Would the farmer understand if he found a dead calf with its throat cut? I also questioned if I actually had what it takes to kill a calf. The answer was clearly no, especially not with a small blunt penknife. My dilemma was futile, as in the meantime the calf sadly stopped breathing and slipped away. I had seen a lot of dead animals on this trip already but this was the first time I helplessly watched something die. Before I could resume my journey, I had to spend a few moments processing what I had just seen.

Finally, the seaside town of Melaque emerged at the bottom of a long hill and I mustered up enough determination to limp into town. On reflection, it was probably the state of my running shoes that had exacerbated the knee injury, as I had purchased one pair some 2,000km before in San Diego and the other 800km away in Guerrero Negro. Proper, authentic running shoes were still hard to come by and expensive when you did find them. I needed to resolve this as soon as possible.

Military roadblocks are normal in Mexico, especially as you approach larger towns. No matter how accustomed you become to them, they still inspire fear, especially as the Americans I'd met had warned me of the risks. So as I approached, I started to worry they were going to take exception to a foreigner running with no supervision or support but instead, I was pleasantly surprised. Six soldiers, dressed in camouflage with bulletproof vests and large machine guns were waiting at the checkpoint. They looked mildly threatening as they ushered me over and I tensed in anticipation but their stern looks melted into smiles when they learnt what I was doing. Soldiers gathered round my stroller with genuine interest and my fear dissipated as the senior officer reached for my straw cowboy hat and asked to have his photo taken pushing my stroller. Snapping away, we joked around and a friend of the soldiers turned up with a tiny puppy to photograph on the stroller.

A larger town was awaiting me further along Highway 200. Manzanillo's size could make it a blessing and a curse. In this instance, a pleasant surprise as my mother had been following the tough time and injuries I'd had of late and decided to book me into a fancy hotel as a late Christmas present. Along the highway I'd seen a billboard for the Camino Real and cheekily tweeted them asking for a free night's accommodation. Much to their credit, despite not giving a free room, they did offer a discount.

On arrival I immediately felt out of place among the other clientele. To these wealthy businesspeople, the sight of a filthy traveller in the lobby must have seemed perplexing. Still my room was amazing, with stunning views over the swimming pool and the ocean beyond. Everything was perfect, except I couldn't relax and when I looked at the price of everything on offer I couldn't imagine spending that much. That evening, I went in search of somewhere more in keeping with my current situation, selecting a small restaurant where I could have a beer and eat chicken wings. It seemed that the more time I spent on the road the less comfortable I felt with the more refined things in life. Instead, I liked to blend in with my surroundings.

My knee was still a concern, so I resolved to stay in Manzanillo another night to recover. As strange as it sounds, I couldn't stand the luxury hotel for another night and decided to cross town, which would also make it easier to leave the next day. With tentative steps, I passed the luxury coastal homes and large hotels before cutting through the city to the quieter, more residential area. Using TripAdvisor, I scoured the list for the cheapest hotel I could find. Once through the door, the owner guided me through a warren of low-lit green corridors and into a room with no windows, a single bed and an open-ended pipe that punctured the wall to form the shower. It may reveal something of my mental state at this point but I immediately felt at home and physically relaxed as the door shut behind me.

With a rested knee and feeling sufficiently strong for the next stage of my adventure, I took to the road once again wearing a new pair of running shoes from a proper sports shop I had found the day before. I should have bought two pairs but in truth I could only just afford one! There were two ways out of town—the toll road or the local road. Experience told me that pedestrians weren't always welcome on toll roads, so I opted for the steeper local road over a headland and back down to the agricultural area below.

At a junction two large and slightly intimidating policemen were waiting. Once again, I feared the worst. On explaining what I was doing and where I was going, their demeanour changed dramatically. Why was I was taking the harder route? Without hesitation they insisted that I divert to the more direct dual carriageway where I would be safer with a wider hard shoulder. I required little more persuasion and set off, before soon reaching Cuyutlán, a charming seaside resort with a laid-back holiday vibe and black sand beaches. Here, a quaint little hotel on the main road had a small room available, not to mention delicious food and even a craft beer.

Heading via Tecoman to San Juan de Alima and onwards to La Playa Sola, the temperature started to soar and as the road moved inland the sea breeze all but disappeared. The combination of heat and hills started to sap my energy, so I opted for a detour back to the coast to camp. A road sign promised a beach resort but when I got there very little had actually been developed. These sites seemed quite common along the coast of Mexico as investment in numerous developments dried up as tourist numbers fell.

At a small shop on the side of the road I stocked up for a night camping on the beach. The spot I eventually chose reminded me of the stunning beach at the end of The Shawshank Redemption. Pure golden sand sloped down to the sea with boulders marking the headlands at each side of the bay. I found a small rundown palapa, tied my hammock between two of the struts and fell asleep to the sound of the ocean lapping against the shore. While resting, I was visited by a local man who lived just up the coast. 'Sir, would you like to eat fresh fish tonight? Maybe even lobster?' he asked. The thought of fresh food sounded tempting as I didn't have anything to cook. 'How much?' I enquired.
'I be back soon to tell you,' he shouted as he disappeared back to his house.

It turned out that there was just one lobster. Lying in my hammock contemplating missing the chance to eat lobster was too much to bear, so I agreed and was instructed to make my way up to his house later that evening.

The experience that followed was as awkward as it was special. The hut was essentially one large open space with numerous hammocks hanging from the beams. In the warm kitchen a big log fire also acted as the cooking space. Men lay in the hammocks while the ladies pottered around the kitchen making tortillas. Right in the middle of the room was a table with one place setting. What could I do but awkwardly take my place at the table? I was acutely aware of several pairs of eyes following me. Like a laird in a period film, I was presented with a delicious-looking lobster served with rice. I ate each mouthful under the watchful eyes of my hosts, paid and retreated back to my hammock for the night.

The next morning my daily routine of checking my kit revealed that part of my stroller had cleanly snapped off the frame leaving the structure less rigid than it should be. There was little I could do while on a beach except rearrange as much stuff as possible to make it more stable and go in search of a mechanic.

A couple of kilometres along the highway I found a small tin-roofed mechanics' yard just off the main road with a few guys milling around outside. They greeted me warmly and were immediately keen to help with my strange request. They abandoned what they'd been working on to assess the damage and came up with a far more sensible solution than anything I could have proposed. I was led into the office and even offered fresh ceviche while the welder got to work. Within five minutes the stroller was fixed and ready to hit the road again. It is worth noting that this little adjustment made it to Buenos Aires and still holds the stroller together today.

Not long after leaving the mechanics was a sign for a turtle sanctuary. Curiosity, as well as the heat of the day, got the better of me and I decided to check it out, discovering that it was possible to volunteer. This meant that for a few dollars you could use the bunkhouse and help with the collection and release of the baby turtles during the night. This was an opportunity I could not miss.

After sheltering in the bunkhouse from the heat during the day, when the evening finally arrived, I was taken to the volunteer meeting. Most were locals who gave up their time to help and everything was communicated in a very rapid local accent that I barely understood. In the end, I just went where I was told to go and was partnered with an older guy. When the time came, we walked out to where the turtle eggs were buried. My companion switched on a bright light attached to a pole above an orange bucket buried in the sand, laid down on his mat and promptly fell asleep. I tried to follow suit but was too excited to snooze. As the night rolled on, not much happened other than my partner deciding to get up and leave without uttering a word. Lying there next to the lamp, I had no clue what I was supposed to be doing. Then, in the shadows, I noticed a small rustle and a small grey, baby turtle ventured towards the light. Just before reaching it he fell into the bucket and was unable to scramble up the slippery sides. Over the next hour or so the march of the turtles continued and soon hundreds had appeared and quickly filled the buckets.

I later learned that the baby turtle thinks it is heading for the light of the moon and ultimately the Pacific Ocean. All these babies had hatched from eggs laid by adult turtles a couple of months before and had been relocated by volunteers to a part of the beach they could easily monitor. This allowed them to protect the hatchlings from the birds that hover overhead at hatching time, waiting to pick them off one by one as they scuttle across the sand.

After a while, a fellow volunteer arrived and informed me that it was time to release them into the ocean. We'd collected roughly 200 newly-hatched turtles at this point and together made our way to where the waves crashed against the beach. This was the crucial moment as we had to get the turtles to the sea as quickly as possible without the predatory birds feasting too much first. My input was for just one evening but there was a huge sense of satisfaction about boosting these baby turtles' prospects for survival.

On our way back to the bunkhouse, we stumbled across a large female turtle digging her hole to lay the next batch of eggs. It was immensely rewarding to watch this huge female go through all this effort for her children. A perfect end to an amazing night.

Although a little tired from the late-night turtle watch, I was full of positive vibes when I started running the next day. By lunch I was famished, so ordered some meat, rice and beans at a small roadside restaurant. While enjoying my drink, a group of young men turned up carrying machine guns and rifles like a local militia. My previous self would have been intimidated but it seemed I was getting used to the number of guns on display in Mexico. So many people on my journey thus far had carried guns: the police, military checkpoints, security guards and even mall guards. I learnt to judge the safety of a situation by the temperament of people carrying them.

In this case the locals were unfazed and I figured that if it was OK for them then I shouldn't worry either. I couldn't allow myself to be scared all the time.

That afternoon I made it to a small community called Chocola and the village at the end of a rough dirt track seemed deserted. Nearly all the children were still at school and most of the adults were out of sight. Thankfully I was able to resupply on water and on my way down to the beach to camp, a small boy tagged along. Despite our inability to converse, he stuck around for a while just playing on his tablet and watching my every movement. It seemed he just liked having some company and to be honest, so did I.

Zihuatanejo was still 250km further down the road, but after a night sleeping on the floor of a mechanics' workshop the night before and only being able to cover 20km in a day I knew I needed a rest day. I decided to leave the main road about midday and follow a more interesting-looking road that stayed closer to the ocean. There stood a brightly-coloured tree, beneath which a sign painted in bright blues, yellows and reds stated: 'Magic Tree—tell me your problems and I will resolve them'. At that moment all I wanted was somewhere to rest.

Las Peñas had a relaxed, slightly hippy-ish vibe. The road was lined with blue-painted stones below small stands depicting brightly coloured paintings of turtles swimming and children holding their necks. As it was low season finding somewhere to stay proved tricky as the only open hotels seemed slightly upmarket and therefore out of my budget. Rather than give up I decided to try my luck and see if I could negotiate a cheaper rate. The first hotel flatly turned me away, but undeterred I walked into the next hotel despite not feeling amazingly confident in my tactic. The lady at the reception desk of Hotel Gambusino was more receptive to my case and as it happened, the owner was in the next room.

He was everything you wanted from a Mexican hotel owner. Although tall and imposing, it was not in a threatening way. He looked travelled and a cigar rested in his hand. When he invited me to join him I duly answered the series of questions he asked. He reclined in his chair and thought for a second then leaned forward.

'I'll tell you what I am going to do. You can stay here in our hotel for two nights and you don't have to spend a penny. Just relax and enjoy your rest.' I was blown away by his generosity and when I saw the room I'd been allocated, I couldn't believe my luck. The next day I found a hammock on the beach just around the headland from the hotel, bought a Modela beer and stayed there all day just watching the world go by. Before leaving this enchanted cove I realised that while I needed these rest days to recharge physically, I just spent the whole time thinking about the road ahead.

After the more open coastline of Las Peñas I stopped in Playa Azul for lunch before pushing on towards Lazaro, an industrial town with a huge ArcelorMittal refinery on the outskirts. Traffic here was dominated by big lorries and four-wheel-drive trucks, in stark contrast to the seaside towns of recent weeks. Zihuatanejo was my focus, so stopping here was nothing more than functional. I found a small hotel in the city centre and decided to give the stroller a proper service in a real bike shop. That evening I chilled by eating delicious street food and drinking beer, smiling inwardly at being the only gringo around.

The pitstop did the trick, as the next morning I awoke full of energy and raring to go, relieved that my wheels had been serviced. It was now just 100km to Zihuantanejo. The good vibes lasted all of five kilometres until I noticed that one of my wheels was shaking more than usual. It transpired that the mechanic hadn't fixed the wheel back properly and the ball bearings were at risk of falling out. My only option was to retrace the 5km back to the shop. To say I was angry and frustrated would be an understatement, as when you are targeting hefty distances and specific destinations you don't need to add additional kilometres. At times it was hard to process these emotions with the weight of my mission bearing down on me. Luckily, as was normally the case, this bad experience was mitigated by meeting more friendly policemen who seemed genuinely interested in what I was doing. They stocked me up with much-needed fresh water, too. Little things like that continued to happen and they restored the positive frame of mind I needed to keep going.

Importantly, I also had a couple of interactions with people back home. First, while pausing at a coconut seller I received an email from Jake, a family friend, saying:

'Thinking of you and hope all going well. No doubt your beard is still growing so be careful you do not trip over it. Looking forward to seeing you at Melissa's wedding. It will be a bit strange for you to not be moving all the time but perhaps you will make up for it on the dance floor!!!!

Anyway, continue to have fun and keep covering the miles.'

Later that afternoon my good friend Al called and as I ran we discussed the change in my life and his desire to do something similar. It's amazing to think back now because after that call Al did indeed take a sabbatical and later we would climb Aconcagua together, as well as trekking part of the Hayduke Trail.

Boosted by the warmth of familiar connections, I ventured to a bar that evening and met a local chap. After chatting, he offered a place to camp at his house and a shower. Overjoyed, I followed him down the back streets to his house where he showed me to a patch of flat, if a little rocky, ground. His house had a walled-off garden, where he indicated a barrel of water and bucket. He told me I could get naked, and he would make sure his wife and family were out... I got on with it!

CHAPTER 11 - CLOWNS AND INJURIES

DISTANCE TO BUENOS AIRES: 11,811KM

My reason for visiting Zihuatanejo was because it is the beach Andy Dufresne headed to at the end of The Shawshank Redemption. Little did I know that during a testing 60km stretch, I would witness something even more disturbing than the dying calf or killing of an opossum. Redemption from this would have been very welcome indeed. Rounding a corner, I spotted two horses and a foal lying in the road. The horses were dead and had been for a few hours. However, the foal was alive and had suffered horrendous injuries. As a keen horse rider, this was particularly hard to witness. I asked a local what had happened and he told me a lorry had hit them in the middle of the night. It shocked me how little anyone seemed to care. With no way of dealing with the situation, I set about looking for someone who could. First, some road workers on the side of the road had machetes, so I asked whether one of them could deal with it. One of their number point-blank refused, offering instead to lend me the tools so I could. That was when I ashamedly realised I just couldn't bring myself to do it. Next, I found a policeman and explained the situation, asking whether he could shoot the foal. This too was refused, as he said policemen firing guns could cause a massive issue, although he assured me that he would make sure it was dealt with. Feeling quite helpless, I struggled to come to terms with my response and tried very hard to convince myself there was little I could have done.

When I got to Zihuatanejo I found a bustling town full of activity and tourists. It was beautiful in a manic way, but I was so drained I ended up walking around aimlessly trying to find somewhere to sleep. I finally struck up a conversation with a resident American who took me to a hotel. Leaving the hotel for dinner was a struggle, but on the way back from the town centre I heard a girl's voice call, 'Jamie!'

I thought I was going mad because there was no way anyone here knew me and especially not an English-speaker.

But then there was that South African accent and when I turned around I saw Jamey, the girl I'd spent my birthday with in San Diego. To say I was happy to see her would be a crazy understatement and my spirits instantly lifted. Among many mutually heartfelt interactions, this was the person that I had most deeply connected with and to see her after such a gruelling day was just what I needed. We joined her new work friends from the superyacht. Fun had not strongly featured over the last few hundred kilometres, but this made up for all that. We drank and chatted and then Jamey came back to my hotel. However much I prevaricated between craving time alone and bemoaning my solitude, that night it was so nice not to be sleeping alone for a change—exhausted or not...

The next morning we breakfasted together before meeting the rest of the crew at the beach. It was hard to fathom that yesterday I was running along the roads alone and now here was a completely different world, hanging out on a superyacht, paddle boarding and enjoying the finer things in life. On days like these I could completely forget my quest to run the Americas and indulge instead in real life.

Looking back, I could have stayed longer. I was making great time and they weren't going anywhere. A break to soak up as much normality as possible would have been good, but that insistent drive to keep pushing wouldn't go away.

In the small town of Petatlán, the last room in the only hotel had been taken by a cyclist, apparently an American, but the owner took me to a deserted building where I could set up base. The identity of this mysterious cyclist would preoccupy me for days of empty landscape.

At Zacualpan (not to be confused with the larger inland town and shrine of the same name), some locals invited me to camp in their garden, which was deemed a little safer than the village. When everything was set up, we sat on the porch and chatted. These people were intrigued by how I got into the United States. One had previously made the illegal journey across the Mexican border. He described how much water needed to be carried, how hard it was and how the authorities wait for groups to get over to America before swooping to catch those who organise it Stateside. It's not known exactly how many people die attempting the journey, but the International Organization for Migration estimated it in the high hundreds in 2022 alone. To my hosts, it was incredible that I could just arrive in the US and not be questioned. Even in 2015, Mexican nationals had to apply for a visa and provide evidence of a certain amount of money in their bank account (I think it was $11,000 at the time). And it would be even more stringent now. The man who had illegally made it to America had had a baby there and wanted to see his family again. It came across that wanting to be in the US was just to get a job, work hard and have the benefits of schools and hospitals. They just wanted better lives.

I was made very aware that running alone here was a little crazy and that I should not stray too far from the road. Talk turned to the small communities in the hills where the drug lords exert complete control. Everyone worked in something related to drugs, the police were too scared to enter, and people were too dependent on their meagre wages to stand up to it. Interactions like this give a lot of food for thought and when you spend so much time alone, your brain mulls over things for hours on end. It's glib to say I am lucky to have been brought up in the UK but sometimes I do feel gratitude for where I was born.

Thankfully I had no encounter with danger beyond these warnings, and safely reached Coyuca de Benitez, a quaint little town with an amazingly busy market in its heart. Entering the town required crossing a wide, lazy river with restaurants lining the east bank, while on the opposite side of the bridge families gathered to bathe and clean their clothes in the water. To access the restaurants meant carefully walking across the thin wooden planks of a makeshift bridge. This strange structure stretched for 50–60 metres across the slow running river but was so narrow you had to squeeze past people coming from the opposite direction. Once on the other side, the restaurants' colourful plastic seats cheerfully clashed, while the huge menus tantalised.

Pressing on, the sprawling Acapulco would offer my next rest and its reputation for old-world glitz and glamour was hard to ignore. It had come to prominence as a popular tourist draw in the post-war 1940s through to the 1960s as a getaway for Hollywood stars and millionaires. What I didn't know, and I am glad I didn't, is that due to a massive upsurge in gang violence and murder since 2014 it has been one of the deadliest cities in Mexico and top ten in the world, and the US government has warned its citizens not to travel there. In 2016, the year after I stayed, there were 918 murders, and the homicide rate was one of the highest in the world. Facts like this are part of the reason I tended not to research the places I was running through. My route was pre-determined and knowing these kinds of details wasn't going to help my mission. Maybe sometimes, ignorance is bliss, though some may argue naivety.

On the day I ran towards Acapulco, by some miracle the only damage I sustained was that my stroller had two broken spokes and two punctures, a broken pump, my phone battery died and I ran out of credit on my phone. This catalogue of disasters led to a day of running fuelled by frustration, but I thank my lucky stars that was the extent of it. Without full capacity for either mobility or communications, I felt extremely vulnerable. When I finally arrived in the city, the hotel I had chosen wouldn't give me a room at the price that was advertised online, so full of determination I made it my mission to get the room at the price I wanted, even if it meant buying a McDonald's to get free Wi-Fi while being pestered by Jehovah's Witnesses and seriously irritating the person at front reception. That night, I went out into the town and tried to find people to drown my frustration in.

Once again, I found myself conflicted. Acapulco had been a magnet as I ran, somewhere I envisaged interacting with people and enjoying comforts but once again what I thought I wanted and what I actually needed were so disconnected. The sheer volume of people, the chaos and the cars made me tense. Miserable, I ended up just sitting in my hotel room and drinking a six-pack of beer alone. An alarm bell went off in my subconscious but I ignored it for now.

The road after Acapulco was sometimes hilly, sometimes not, sometimes hugging the coast and at other times bending inland and military checkpoints gradually built-up along the way. It was on my way to the small town of San Marcos that I heard another shout of 'Jamie!' Sadly, this time it didn't come from a beautiful South African girl but an American voice I vaguely recognised. It was Brooks—the American cyclist who had cooked us all spaghetti in the Redwood Forest all those months and kilometres ago. Seeing a friendly face in the middle of nowhere was a delight, and it turned out that it was he who was the mysterious cyclist that had taken the last hotel room in Petatlán a week ago. We briefly chatted on the road and decided to take a day's rest together, both craving some connection between like-minded spirits. When I got into San Marcos, Brooks had already sorted out the hotel accommodation for the next two nights, and even paid for me. We had a great night catching up in a little restaurant next to the hotel where an old lady prepared some delicious home-cooked food.

The next day was devoted to walking around San Marcos, exploring the maze of tight backstreets and bars covered by colourful awnings. Here we discovered the Michelada, the Mexican bloody Mary, made with lager, clam juice and Bloody Mary seasoning. We gawped at the rows of dried fish laid out of tables in the market, trying to ignore the sweet stench of it all. What really stood out was how different Brooks and I were when it came to interaction. I tended to mostly observe, but Brooks (being American maybe) was far more interactive and had no compunction about walking up to people and speaking English. Initially this made me feel a little uncomfortable, but Brooks is such a lovely, kind chap that no one could see his overtures as anything but genuine curiosity.

Of course, hanging out with Brooks was great fun, but even more rewarding was being able to share the experiences with someone who understood the significance.

The next few evenings were spent together, but during the days we would split up due to our slightly different speeds, though I might add our overall distances weren't too dissimilar. Rolling into a small roadside restaurant for lunch, a smartly dressed man invited me to sit down. When I asked for the menu, he led me into the kitchen and started to lift the lids on bubbling pots of stews and soups. There were chicken, vegetables and fish dishes but also one pot that he didn't open. I asked what it was and he duly lifted the lid to reveal a red sauce with dark chunks of meat. The skin that was visible was clearly reptile. 'It's iguana,' he said, 'would you like to try?'

Slightly torn as to whether it was ethical to eat iguanas, my intrigue got the better of me and I was soon presented with a steaming plate of admittedly delicious-looking stew, complete with chunks of legs and tail. After meticulously peeling off the skin and prying away the white meat it tasted surprisingly good. The taste is, of course, that of chicken but the texture resembles a meaty white fish. Suitably nourished, I pushed on to re-join Brooks again that evening.

Marquelia, 40km down the road, was to be our last night together, for now, and once again Brooks excelled at finding rooms in a cool little hostel. The town itself seemed to have an unusual number of men dressed in feminine clothing with lots of make-up, nail polish and many businesses were run by outwardly gay men. We learnt that families don't make an issue because if a son is a transvestite or gay then he's more likely to stay at home and look after the parents rather than marrying and moving away as a straight child might.

Running straight into a ditch on the road after leaving Marquelia wasn't my finest hour, but then again, I did have two massive dogs chasing me. I didn't catch more than a glimpse of them, but they seemed huge, their massive open jaws full of alarmingly pointed teeth. I broke into a sprint, lost control of the stroller and careered into a ditch. Thankfully I must have left their territory as they lost interest at that point, and nothing was injured except for my pride. Dogs in Mexico frequently run out of nowhere or bark from roof tops and behind wire fences. Normally it's merely territorial but that still doesn't stop it from being mildly terrifying.

Valentine's Day is a big deal in Mexico and it coincided with my visit to a town called Santiago Jamiltepec in the Oaxaca region. There, groups of young people were dressed up in brightly coloured masks and school uniforms for a fiesta that brought the market streets to life. Below my hotel window, a parade with dancers in pairs, the ladies in bright colours and the men all in white with red scarves around their necks, dancing to traditional Mexican music played on trumpets and drums and all led by cowboys on fine stallions. Plastic tarpaulins stretched over the back streets, where stalls were crammed together offering absolutely everything from ripped-off CDs of Mexican music, cooking equipment, food, clothes, farming products and trinkets. So many people had gravitated to the town for the fiesta, ranging from local farmers to the evidently wealthy. In the town square a stage was set for a regional dance competition and demonstration, which I watched in admiration. So much positivity, vibrancy and colour was on display that evening. The dancing ranged from couples in line with the men all in white and wearing straw hats, while the ladies wore wonderful whirling dresses that kicked up as they twirled. A highlight was watching the men performing the Danza de la Pluma, with their huge, plumed headdresses. This dance tells the story of the Spanish Conquest from the indigenous point of view and is performed with vigour and meticulous precision. After the dancing had ended, I persuaded one of the dancers to let me try on his three-foot tall, white, red and yellow feathered creation, and we both beamed at the spectacle of it paired with my running clothes.

The rest and distraction of the fiesta meant I was back on good form. Even better, the scenery was lush, with wide easy-flowing rivers and fields in which workers could be seen picking and packing papaya fruit. By now the road had veered substantially inland. People were friendly on this stretch, happy to break off work for a chat or even offer me a glass of cold Coca-Cola. Not far from here was a small beach village I'd read about called Playa Roca Blanca, and I was overdue a small detour. After a bumpy local road, the village was very small but quaint, more a collection of restaurants on the beach, where I booked a small hut behind one of them for the night. Saying it was rustic would be an understatement, but it had a bed and the lady who ran it was very kind. I spent the afternoon relaxing on the beach and got acquainted with a few of the local expats.

A highlight of Mexico was my Forrest Gump moment on the approach to Puerto Escondido. Not long before evening I had stopped to buy some coconut water and was getting ready to run again when 10 school kids started running with me. At first I thought they were joking around but soon realised that some were being serious. Even though they wore full school uniform, including leather shoes, and carried backpacks full of books, their turn of pace was surprising. I felt a little conscious that people might question a European man running with a group of random local children but none of these concerns seemed to be shared by the kids. The group slowly whittled down to six and then four. After five kilometres we stopped at another stall, where I bought my running companions refreshments. Later, we arrived in the town, and they assiduously guided me to the street I needed to be at before we all shook hands and they disappeared home, smiling broadly. It was at times like these I remembered how alien I must have seemed in some of the rural places. It was an incredible experience for me and filled me with huge positivity.

By now, far less of Mexico lay ahead of me than I had already covered and I finally got around to buying a ticket home for my sister's wedding. I planned to leave Panama City on 19th June, which meant I had to cover 3,100km in 117 days. To get there I would have to run to the end of Mexico and then the entire length of Guatemala, El Salvador, Honduras, Nicaragua and Costa Rica before arriving in Panama. Just trying to process all that was hard enough but a deadline was a strange new addition, even though it wasn't that demanding from a mileage point of view. There was no way to gauge whether I'd given myself too much time to make the journey to the airport or not enough. Nevertheless, just thinking of being at my sister's wedding gave me the drive to keep going.

The scenery was becoming even more tropical as the coastline curved east towards the border. I had decided to stretch my distance to 60km, aiming for Mazunte, a small hippy village on the coast. When I arrived, most people were bewildered to see a hairy runner pushing a baby stroller but even more so when I ran directly to a small shop, bought a beer and collapsed onto the pavement. That night I was glad of a small makeshift room tucked behind a restaurant on the beach, luxuriating in the sand around my bare feet.

It was tricky to strike the right balance between making progress and absorbing the local ambiance. I continually struggled to relax and linger anywhere due to a fear of losing my sense of purpose and direction. In my mind, the run was my work and therefore always took priority over everything else. I had to keep reminding myself that I wasn't a traveller or backpacker, but an adventure runner on a mission. That may be one of the reasons I never really developed deep connections with the people I met. With hindsight, I regret this but realise it was this focus and determination that ultimately kept me going day after day, regardless of the pain, the loneliness or the struggle.

My coastal run took me past the liberal beach of Playa Zipolite, where nudists casually walked by the restaurants, and then to Puerto Ángel, a small port village built around a cove. While passing through La Crucecita, I heard of a small surf village called Barra de La Cruz and needed no persuasion to take a couple of days' rest to unwind. Most of my rest days were spent catching up with admin in large towns rather than switching off. After 195 days on the road and firmly in the groove on the running side, dealing with countless small incidents had built up fatigue and was making me less able to handle new issues.

As I neared the turning for Barra De La Cruz, a cyclist behind me called out. Half thinking it was going to be Brooks, I realised this topless rider wasn't him. It was Dominic, a German cyclist who was touring Mexico. We looked at the map and saw a small restaurant a few kilometres further on and agreed to meet there. On the long hill that followed I managed to overtake my new cycling buddy but was quickly put back in my place on the descent. When I finally caught up, we ordered a litre bottle of Corona each, over which to share stories. Like me, most cyclists I met along the way seemed to be consciously journeying from one place to another, but Dominic had decided to spend six months cycling about 12,000km around Mexico with no particular direction in mind. He told me that he wanted to submerge himself in the culture rather than just cutting through a country to say that he'd completed a certain distance. Even though this was exactly what I was doing, I could understand his freewheeling approach and was a little envious.

Quickly finishing our first beer, we moved on to a second before attending to where we would stay that night. Dominic was having a cash flow issue due to the lack of ATMs but wanted to go to Pepe's Cabañas, where I had been recommended to stay. Here was the perfect opportunity to pass on the $20 Jim had given me in the Redwood Forest, so I gladly offered it to Dominic, explaining that he would, in turn, have to do the same. I often wonder if he did.

The resort was tailored for the surfers who gather here during the summer but as it was out of season, we pretty much had the place to ourselves. As a result, it was the perfect place to switch off and recharge my mind. My accommodation was a small, round hut with a bed and mosquito net and other than the kitchen there was very little going on. It was a mere 10-minute walk from the beach, whose golden sands gilded the shoreline as they stretched down the coast.

That night we sat in a local bar where Dominic got chatting with the owner. I was jealous of how good his language skills were and was embarrassed at how limited my own Spanish was. My level of competence allowed me to ask for hotels, order food, and so on, but I had not managed to achieve a conversational stage, and this was preventing a deeper interaction or full experience of the places I visited. Many people told me that I should spend more time learning but running 30–60km a day meant I was both mentally and physically exhausted at the end of each day, so in my solitude I spent very little time speaking to people.

Dominic left early the next morning leaving me all alone at Pepe's. The place was absolute heaven, and the perfect chance to do my washing, go for walks on the beach, do yoga and catch up on admin. I learnt that Barra de la Cruz was a really tightknit community that had struggled to keep the large corporations away. It was run by local indigenous people who deemed it important to keep it that way, despite the considerable money that could be made from selling out.

My next target was 250km away, in a small village called Zanatepec. One of my dearest friends, Suse had, unbeknownst to me, been tirelessly contacting running brands to find me some new kit. She had worked with one of her colleagues, Lindsey, and convinced Nike to send a care package to Mexico. The logistics had been very complicated because we never knew where I would be at any one moment or where a package could be sent. With perseverance, we managed to come up with a plan that would see the package delivered to a Warmshowers host, who would hold it until my arrival.

This little gift gave me something to focus my efforts on and the drive to keep going. The days were getting pretty long and hard as there was very little to do other than run, meaning I was averaging over a marathon a day in an attempt to keep sane. I passed through Santiago Astata where another fiesta was in full swing and then onto Salina Cruz. This town is home to a naval base, made interesting by the infrequent glimpses of its former glory in the decaying architecture that now appeared sad in comparison to the modern hustle and bustle. Finding somewhere to sleep proved tricky and while trawling the streets, a guy with tattoos all over his face walked up and just stared at me. Somewhat daunted and with the next town only 10km further down the road, I decided to push on.

Before getting very far, my unusual demeanour had clearly piqued the interest of a policeman, who was chilling at the side of the road. He asked what I was doing and when I explained he reached into his pocket, pulled out some pesos and told me to get some food at the next restaurant. A little further along, a car pulled over and a father and his son got out. The young kid, obviously returning from a sporting event, came over and handed me an energy drink. These little acts of kindness spurred me on.

The next town, Tehuantepec, was traditional and rustic and I immediately found a charming hostel. I got chatting with the host and we decided to pop out for food together. Just as we were leaving, my foot found some uneven pavement and I went over on my ankle. This was my 200th day and I had successfully covered 6,250km already with nothing more than niggles, so how could this have happened? Unable to walk, my fear was being unable to recover as quickly as I needed.

Despite the swelling and the dark blue bruising starting to spread like ink around my lower calf, I was determined to keep it moving. In defiance I went out for dinner that evening with a couple of girls from the hostel and my mind was briefly taken off my ankle.

As we strolled, me limping, we were drawn to a crowd of people in the town square who were entranced by some entertainment. As we got closer, we found a clown performing. Immediately on spotting three foreigners in the crowd, he seized the opportunity to get the crowd worked up and encouraged them to start shouting 'Gringos' at us. Although in jest, it was daunting nonetheless. The clown decided to get us up in front of everyone to act out a play, much to the amusement of the surrounding audience. My Spanish was not up to understanding his quick speech or humour, but even I could gather that they were making fun of the fact that we were not Mexican. It was a surreal situation in which to find ourselves, especially as we didn't even know each other. However, in good humour, we took the abuse even though it did leave a bitter taste in my mouth, especially as I was already down about my ankle. But what choice did we have?

Running the next day was out of the question. I was bedbound as my foot was now far more swollen and putting weight on it was excruciating. The only time I left my room was to order a large alfalfa drink from the café next door, as it came in a plastic bag with lots of ice that substituted for an ice pack.

The next day saw a little improvement, so I contented myself with hobbling around to explore the town a little. My annoyance at being stuck was only partially mitigated by the fact that the hostel was nice and the village undoubtedly had a charm to it. Not enough though, to prevent me from upping my efforts to recover as quickly as possible through a mixture of ibuprofen, icing, rest, elevation and movement.

This was partially successful, and despite the pain I somehow managed to cover 56km after waking the next morning and heading for Zanatepec. My Warmshowers host and care package holder was Rodrigo, who turned out to be a great guy with an awesome family and an enviable property. As luck would have it, a group of cyclists had also stopped there during an adventure they called 'Pedal South' and who were making a documentary about their travels. One was a web designer, another a filmmaker, and a couple of photographers. We all bunked up for the night in the courtyard, with our sleeping mats on the ground or hammocks tied between trees.

The package that had arrived opened to reveal some superb Nike running kit. Since the beginning of my adventure, my running shoes had been whatever I could get my hands on and normally the cheapest available. Now I had three brand new pairs of shoes, a running watch and a set of running clothing. Before heading off, I gave Rodrigo some of my old clothes to donate to the local school as well as one of my running shirts as a thank you for everything he had done.

After 127 days and nearly 3,700km in Mexico, I finally crossed into Chiapas, the last state before Guatemala, the fourth country of my adventure. I'm still not sure if it was the excitement of a new country or a desire to find a television to watch the Calcutta Cup that powered me on for those last few days in Mexico. Both, probably.

On top of the hill when I got to Arriaga stood a whitewashed church with red edgings and a beautiful dome, decorated with attractive statues. Inspired, I sat and reflected on everything that had happened so far, what I had achieved and the prospect of getting to Panama in time for my flight. The sheer size of the USA and Mexico had made them epic challenges to tackle, which made the next countries seem somehow manageable, especially as they might offer ever-changing cultures, new and exciting challenges and unknown adventures.

Heading south once more, Tonalá gave me an opportunity to experience Mexico as I wanted to remember it. An old colonial building fronted the main square, and in its heyday it would have been stunning, even though by now it had given in to decrepitude and become a tired yet charming hotel. Its elderly lady owner was confined to the room on the ground floor and children scampered around the courtyard. In the evening the town square came alive with bands, dancing and churros stalls as different generations all congregated together. In a restaurant that overlooked the scene, I enjoyed the atmosphere, albeit with a tinge of sadness that British people don't seem to embrace a lifestyle where families come together in one big open space. In England—at least in part as a function of frequently inclement weather—a lot of celebrations tend to be inside, behind closed doors.

On the way into Tapachula, a final military checkpoint straddled the road and while most of the soldiers were friendly, one seemed especially interested in Carlos, the teddy bear I'd rescued from the side of the road in Baja. He squeezed him every which way, presumably checking there were no drugs hidden inside. I got a little worried that I was going to be in for the same kind of treatment when the soldier pulled on a plastic glove but, to my relief, promptly removed it!

In Tapachula, Mexico's main border city with Guatamala, my hotel had strong enough Wi-Fi to allow me to watch Scotland being beaten 25-13 by England at rugby. While I commiserated in my hotel room with a few beers and a packet of Pringles, I had a call from the front desk. It transpired that I'd been summoned as the young night manager wanted to take a photo with me. I obliged and returned to my room. Not long after, he called me back, and this time presented me with a bracelet to remember him by. Things got a little awkward and while flattered by the attention, I politely explained that I needed to rest before scuttling back to my room.

After 131 days in Mexico, and 213 days on the road in total, I had reached my last day in this incredible country. This would be the country I spent the most time in, yet I'd just scratched the surface of this amazing place. Before entering Mexico, I'd heard so many horror stories about the things that could have happened, based on the deep-rooted fears of the people I had met in the US but my experiences here had been nothing but positive. The culture had supported and welcomed me and at no point did I feel threatened or intimidated. Due to my objective to cross the country, perhaps I missed a few opportunities to truly become immersed in the culture and I envied Dominic who had covered four times the distance I had. But I also knew that my journey wasn't just physical, it was also emotional and spiritual. I was already a different person to the man who had left London, but with so much time alone to reflect, I also knew that there was so much more to process and work on.

Leaving Mexico didn't happen without one more drama. Finally arriving at the border necessitated refuelling to enable me to adapt to a new environment and the new set of rules presented by Guatemala. At a small restaurant I ordered my last Mexican meal. But horrible panic engulfed me as I crossed the village towards the border control. Something was clearly awry, as a cold sweat seeped over me and my stomach rumbled. Sometimes you just know a problem is imminent and I started frantically searching for a toilet. Running to the nearest posada I could find, I paid the fee, ditched my fully loaded stroller in the street and just managed to make it to the toilet before my bowels emptied dramatically. This was not how I wanted to finish my time in Mexico.

CHAPTER 12 - CAN I TAKE A SELFIE?

DISTANCE TO BUENOS AIRES: 10,283KM

The last thing a Mexican said to me before leaving was that I had nice eyes, and that was by the immigration officer as she stamped my passport. With a spring in my step, I strolled across the bridge that linked Mexico and Guatemala over the Suchiate River and was immediately confronted with a problem; I didn't have any local currency to purchase an entry visa. As I stood in line trying to explain my predicament, a gentleman behind reached forward and slipped me the money, refusing to accept any repayment. Guatemala instantly filled me with positive feelings, despite all the warnings I had received from the neighbouring Mexicans. Ironic really, given all the warnings I had received about Mexico from the neighbouring Americans.

Once again, the invisible line, or in this case river, that separated the two countries marked a huge change. The architecture was different, the brands on billboards were different, the people even seemed to look different and the foods on offer in the shops were different.

Everything felt a little more basic here, with less developed roads and shops than Mexico, and no locks on the doors at the hostel, but this was all hardly surprising in view of the differing economies between the two countries. Thankfully, the people seemed just as friendly. In a small burger bar, the locals were inquisitive but not threatening at all. Like a gemstone, this adventure seemed to be revealing new and previously hidden aspects to me at each turn.

Returning the next morning to the Pan-American Highway, the number of lorries waiting to cross the border back into Mexico was shocking. The line stretched for at least 10 kilometres, with drivers dozing in hammocks under their trailers or gathering in small groups to drink coffee. A mini economy had sprung up around the traffic jam as motorbikes hovered along the line of lorries like honeybees, selling food and drink at each one before buzzing to the next.

Orientating myself on my first day, I aimed to get to the first big town of Coatepeque, some 34km away, to settle into this new country. While soaking up the scenery, a man on a motorbike pulled up alongside. 'I work for the local newspaper,' he said. 'Can I interview you while you run?'

I had no idea how he knew about me, but it felt great to have an opportunity to share my story, so naturally I obliged. Answering his questions as I ran took its toll on my running, but I tried to humour him. Over the course of the day, he kept returning, trying to eke more information out, and by the end of the day, I was actively trying to evade him.

The approach into the city centre was through a long street market where overloaded stalls competed for space with various live animals on show. The buzz here was palpable and it compounded my positive impression of Guatemala. Coatepeque seemed the ideal place for an introduction to its culture. The wooden-fronted buildings of its packed main square conjured the Wild West of old movies, except festooned with an impossible web of cables and wires. Services on offer ranged from shoe shiners and people selling trinkets to vendors of basic household goods and mobile phone accessories.

Simply being in Guatemala was exciting, but the more I studied the map the more I realised my time here wasn't going to be full of the postcard moments I had dreamed about. A divide seemed to run down the middle of the country with one half hosting the dramatic scenery, ancient ruins and cloud forests, and the other comprising flat agricultural plains that ran along the Pacific Coast. Obviously, as a runner, I had chosen the most direct and flattest path which meant less iconic scenery. This meant accepting my experience of Guatemala would be of the people, rather than the guidebook itinerary.

Large sugarcane factories dominated the skyline, spewing smoke and steam into the atmosphere, and apparently other effluent into the rivers. The waters were grey and even warm and it must have been decimating any wildlife. Clearly this was a sacrifice the country felt it had to make, as Guatemalan agriculture relies on sugar cane, bananas and coffee.

Vowing to one day return to the more spectacular scenery to the north, I pushed on to reach Mazatenango, a large and not especially touristy town. Its beating heart is again a huge covered market where a maze of stalls sells everything from household goods to colourful spices and dried fish. As I was the only gringo in town, everybody stopped and stared, but undeterred, I booked in at a very basic hotel right in the thick of things.

The street my hotel was on sloped down past the market, immediately on my right. There was no grand entrance, just a loading bay with a small opening and a passageway disappearing into the darkness. On entering, I was immediately confronted with a profusion of small stalls, each with a shopkeeper or child at the ready. As I meandered around the maze of commerce and slowly made my way to the centre, the categories of shops morphed in some sort of organised mayhem. Mobile phone accessories merged with fruit stalls and then spices were piled high in large canvas bags and suddenly chairs and tables were on sale. At seafood stalls, little old ladies swatted flies from their fresh fish, crabs and prawns. Rubbish blocked the gutters and there was an ever-present smell of rotting food. There was no conformity here, only confusion. The passageways were tight and the only illumination came from what broke through the cracks in the structure.

The street outside was chaotic with cars, pick-ups, red flashing moto-taxis and minibuses all darting around without any regard for each other, let alone the pedestrians. Nowhere but the pavement was safe, making pavement space understandably scarce. Shops lined both sides of the street—selling pastries, mobile phones, textiles, bike repairs and fried chicken. Large colourful signs poked out everywhere to advertise whatever products were buried beneath. On the frontline outside shops, vendors had set up small tables covered in lollipops and sweets alongside makeshift cobblers and shoe cleaners. On each street corner, ladies in colourful traditional dresses sold fresh mango, pineapple and other exotic fruits, all pre-sliced and in little plastic bags, ready to go. What struck me was that there was no hard sell and none of the harassment or bullying tactics that you might experience elsewhere; here it is one price and that price is for all.

Off the main street revealed a darker side, as security guards leant against the entrances of larger shops, pump-action shotguns at their side. This did not perturb the old ladies dressed in colourful dresses who darted around shoppers, balancing impossible baskets on their heads.

This city is a work in progress, and the bright colours had long faded from previous incarnations of its buildings. A perfect example was the market itself, which still bore the livery of its earlier function as the home of culture and arts. Telephone cables crisscrossed the sky as they stretched in every direction between the tops of buildings that were often unfinished or derelict, though they hinted at a more prosperous past. And pavements were strewn with rubbish from the day, which stray dogs scavenged to seek chicken bones or other edible titbits. Despite the chaos and decay, water trickled down shopfronts in a dignified attempt to keep them clean, and by evening, activity started to die down.

While I initially wanted to wild camp my way through Central America, I decided that it was probably not worth the risk given the advice from locals, not to mention how cheap the accommodation was. Plus my route followed the Pan-American Highway, so there would be plenty of cheap roadside hostels in which to stay.

Selfies seemed to be a big thing here and throughout the day I would constantly be stopped and asked to pose for a photo. The people here had the most amazing smiles, often paired with exceptional generosity. From a free coconut by one family to water from a group of girls, a steady stream of gifts enlivened my days. As I was leaving one town, a family pulled over to ask what I was doing. They seem interested and promise to meet me a little further along the road as they had something they want to give me. Thirty minutes later they caught up and presented me with a wood carving of a coin, the size of a large plate. Although the significance was unclear, the generosity was more than apparent, as I imagined this guy running into his house to take it off his wall. The gift was a rare honour, and after running nearly 3,000km with it, the coin now hangs above my fireplace. I later found out that the coin is known as the Choca and depicts an image of Concepción Ramírez, a peace activist, which is significant for its representation of the country's indigenous people.

As I got closer to Escuintla, a lady pulled over to say that she owned a restaurant and could offer me a bed for the next couple of nights. On receipt of the directions, I made my way there enthused by the prospect of a day off and somewhere comfortable and secure to stay. My room was above the restaurant and provided safe storage for my equipment while I joined the proprietor and her American husband for lunch. We got talking about what to do on my day off and a day trip to Antigua Guatemala seemed enticing. A short local bus journey to the foothills would allow me at least to experience some of the beauty and culture I was missing on the agricultural plains.

Antigua Guatemala is a fantastic colonial town and former capital of Guatemala, so perhaps unsurprisingly, had the most amazing architecture. I spent the day patrolling the streets to visit churches and opal museums. Bereft of the stroller I'd pushed every day for three months, being just another tourist held a strange emptiness.

Tourist itch scratched, my next duty was to write postcards to all the people who had supported my expedition. I perched at the bar of the No Se Café and before long I found myself chatting to the girls who worked there. As time rolled on, the group grew into an impromptu bar crawl, and the discovery that my drinking ability had diminished quite considerably. All thoughts of heading back to

Escuintla that evening seemed to vanish in a haze of alcohol. Looping from bars to hostels, restaurants to bars, somehow we ended up at the first bar, where we stayed into the early hours of the morning. This involved a lot of Mezcal, a Mexican drink distilled from agave, and apparently at one point I declared I was going home though I'm not entirely sure where I intended to go. Everyone warned me that was a bad idea and potentially dangerous but in my belligerent drunken state, I stumbled out into the dark and wandered the streets. At 6am the next morning I awoke, propped up against a wall on a street at the complete opposite end of town to the bus station. This was not a proud moment and had worrying reminders of the life that had led me here in the first place. However, on this occasion I managed to blame the Mezcal.

Being awake at such an early time gave me the chance to walk around the town before it officially woke up, even if I was still mildly the worse for wear. The difference between day and night was quite eye-opening and many homeless people slept in doorways and ATM booths. Outside the local hospital, a line of local people waited, who had travelled to the town to get treatment for their ailments. It seemed a harsh reality to the pretty tourist town I had experienced the day before.

I finally managed to find a bus back to Escuintla and arrived feeling horrific. Despite the hangover, I packed up my belongings and got ready to continue my run. But on pushing my stroller into town I realised there was little hope of making it anywhere that day. Returning to the kind lady and her husband would have been unbearably embarrassing, so I ended up, after a few rejections, in a small run-down hotel on the outskirts of town. The rest of the day was full of regret and self-loathing. To make matters worse, I discovered that I'd left my bank card somewhere in Antigua Guatemala. I was in no state to resolve this and had no option but to continue without it. What I can say with hindsight now is that my adventure was a series of puzzles to solve, even ones like these, of my own making. Had I allowed any one of them to overtake my resolve, there is no way I would have made it to the end.

The next day it was all I could do to set off for Taxisco, where I took a room in a cheap hostel above a funeral parlour. There was a wild feeling to this town, which was remarkable for just how untouched by the western world it seemed. While I did feel a little on edge, I savoured how much more adventurous this place was.

The next day was my last in Guatemala and the running was hard. The roads were up and down, and the temperature was rising again as I dipped below the 10,000km point to my target. Greens were turning to browns, and everything seemed parched. Nearing the border, the positive experiences of Guatemala filled my head and I smiled, before turning my thoughts to country number five, El Salvador.

CHAPTER 13 - RUNS TO A BROTHEL

DISTANCE TO BUENOS AIRES: 9,956KM

You never know quite what to expect when you cross a border but you know the new country will be different, exciting and will definitely require some adjustment. El Salvador was one of the countries I knew the least about and that added an extra thrill.

It was 100 or so metres from my hostel to the border. Immigration buildings in Central America tend not to have clear signs or helpful guides directing you where to go, so you have to rely on your judgement and identify the right people to approach for assistance. Various characters pop out of nowhere trying either to exchange currency or just beg from you. The chap who targeted me on this occasion was on crutches as the lower part of his right leg was missing. He was persistent but not intrusive and I would have given him money if I had any to hand, but one lesson I have learnt is to make sure people don't see you have cash. That just attracts the hordes.

Among those intending to cross the border was a Scottish girl called Ellen, so we decided to join forces for the exit process from Guatemala and together crossed the 400m of no-man's-land to seek entry to El Salvador. Just having this companion amid the confusion made it easier to navigate. With all our paperwork in order, Ellen and I bade each other farewell and I switched into 'new country mode'.

All the maps show the first few kilometres of El Salvador to be completely deserted, but that turned out to not be the case. Compared with the vibrant green of southern Guatemala, now the backdrop was a patchwork of sun-scorched yellow scrub and lush green maize fields, with cows grazing under trees. There was something very tranquil about it. The houses seemed more rustic but generally, it seemed comparable to the rural areas of southern Guatemala.

It was clear I was nearing the town of Cara Sucia because things were becoming more built up and shops started appearing. Everyone was buzzing around trying to sell something or find someone. The first lady I met said 'hi' and when I asked her for directions to the bank, she pointed down the street then shook my hand. This was to pass on the ten-dollar note that was concealed in hers, and I am still undecided whether she was just taking pity on me or that my Spanish was so bad that she thought I was asking for money!

After leaving my bank card in a bar in Guatemala, my cash situation was becoming precarious. My spares were MasterCard and AMEX, which unfortunately were not as widely accepted, so it became imperative to take cash out at every machine that did. Having negotiated my way past the armed guard at the first bank I found, my card was declined. I tried to remain calm and located a second bank to repeat the process. Declined! It was at this point I started to worry. 'Is there any way I can get money out?' I asked the bank manager as I presented him with my cards.

He looked them over, then shook his head, 'I am sorry sir, but to use our machines you need to have Visa or Plus. I can't just give you cash. The best bet is to try the Bank of America.' Relieved to know there was an alternative, I asked where to find it.

'It's about 50km further down the road.'

Dejected and fearing the worst about how to cope in a foreign land with no access to funds, I returned to the street outside. Clearly aware of my plight, a guard called me over and suggested I try one more bank across the street. Nervously, I inserted the card, typed my pin, hit the $100 button and prayed. Then came that lovely revolving sound that announces, 'your money is on the way'. Relief washed over me as I repeated the process as many times as possible. I wanted to go back to thank the guard but decided against advertising myself as a tourist with hundreds of dollars!

Playa Costa Azul, 30km south, was my target that day and as I left town, I stocked up on fruit from a lady on the side of the road. She was sheltering from the sun under a poorly built palm hut accompanied by a lady in a wheelchair and a man who was just hanging around. The fruit was in plastic bags and resting on some ice that was doing a poor job at any sort of refrigerating, but it nonetheless was so welcome, especially after my near drama at the bank.

Arriving dusty and weary at Playa Costa Azul, there were no hotels but some melon vendors informed me that I'd find plenty at the next beach. The day had been long, hard and hot and had required me to tackle lots of administrative obstacles, so it took all my mental and physical strength to put one foot in front of the other for a further 10km. It was as much as I could do to leave the highway and walk the last few kilometres to the coast. I was tired and asphalt had transitioned into a cobbled track covered in black sand, which would have made the going impossible had the sound of the ocean not beckoned. All along this track, the residents craned in the evening sun to see me from their hammocks and waved as I passed. Thankfully a restaurant had a room available, even if it wasn't much more than a shed with a corrugated iron roof. Having covered 150km in just three days, I was desperate to sort my kit, plan the route for the next day and after a quick sandwich and beer, get some rest.

La Libertad, some 100km further south, was the next main town and while I was developing a real dislike of cities, something excited me about this one. To get there, the coast road made me work hard but provided great views and some excellent seafood restaurants. Compared with the towns and cities, I loved the rustic simplicity of the buildings and just how relaxed the pace of life seemed to be. The downside to this is that people had very little, and all along the road, the small makeshift houses were cobbled together from old bits of wood and corrugated iron, from which the inhabitants were trying to sell whatever they could.

On one occasion, when I stopped to buy water, I was approached by a lorry driver who seemed particularly excited. 'You are the runner I saw at the border between Guatemala and Mexico...'

It was hard to fathom that this stranger had seen me in three different countries. We chatted a while to marvel at this strange coincidence and took selfies together before parting company.

Two great overnight spots in particular really enhanced my experience. The first was a small family-run place in Playa Dorada, where I took a day off and even joined the birthday party of the owner. It felt strange to be at such an intimate celebration but also quite warming to be around a family coming together, something I hadn't been able to do for months. The next was El Zonte, a small surf town with hostels and restaurants full of local tourists and even the odd cow wandering aimlessly down the street. Here I wasn't included in the festivities, but it was still cathartic to be around so much happiness.

When I got to La Libertad I discovered a lively, bustling town, particularly in the main tourist area where shops spilled out onto the street. Each one had a colourful umbrella protecting it from the fierce sun, beneath which tourists bought buckets, pots, plates, swimming trunks and so on. Brightly painted, yet ancient buses ran up and down the main streets. After being turned away from the hotels in the popular surfer part of town, I found a sketchy little hostel behind the main strip. Clearly it would once have been elegant, even regal, but by now had fallen into disrepair while the swimming pool was just a hole in the ground, collecting leaves. Scrap metal rusted in every corner, making the whole place resemble a junkyard rather than a hostel. Nevertheless, the room had a bed, a toilet and something that resembled a lock on the door and that is all I needed.

A swim in the ocean seemed a good antidote to this depressing environment, so I headed towards the coast and found a path leading between buildings that teetered on stilts. Instead of sand, the beach was composed of large rocks that had been smoothed by the ocean.

A family played in the sea as people watched from the restaurants above, while I basked in the freedom of swimming and the solitude in an otherwise hectic place.

That night I ate alone in a restaurant full of tourists enjoying each other's company. Once more I felt a cold touch of loneliness because I didn't fit into any group. You might even call it slight jealousy that other people were enjoying themselves and at that moment I had to reiterate to myself that this was more a job than a pleasure trip. There could be no distraction from the task at hand nor could I always expect to enjoy the opportunities around me. I had to have tunnel vision.

Having gone to bed feeling rather down, I woke up with a horrendous stomach problem. I couldn't place what had caused it—the ocean water, the seafood or the rich dinner from the night before. It was even worse than my last day in Mexico and there was no way I was in a state to run that day. Instead, most of the day was spent in the room trying to recover. Desperate for some fresh air, I did eventually manage a brief walk around the town and visited a pier that had been converted into a covered market. On sale were pans for cooking fish, ornaments made out of shells, and of course a huge array of both fresh and dried fish; there were crabs, what looked like sharks and lots of prawns, shrimps and octopus salad, the sight of which did at least take my mind off my discomfort for a while. All the fish was chilled by large chunks of ice melting in huge metal bowls and protected from the flies by ladies waving palm leaves. At the end of the pier, you could watch brightly coloured fishing boats being hoisted up with the crews still on board, much to the delight of the onlooking crowd.

I felt like death and spent the evening bedbound trying to prepare for the next day of running. My tiny room was so hot, had no air-conditioning and only a very basic toilet. This, combined with sweating heavily, being dirty and chronic diarrhoea only compounded my already low mood, and no doubt contributed to my decision to run the next day. Despite not being physically up to the challenge nor able to eat properly and being seriously dehydrated, I was struggling mentally with my confinement, so the only remedy I knew was to keep moving south.

Although a long way from 100 percent, by the next day I felt well enough to risk the move out of La Libertad. My saving grace was that the run was pretty straightforward and for the first section of it, trees provided some shade. The scenery here was truly awe-inspiring between the Pacific Ocean on my right and mountains forming on my left. Crossing a stately river, I saw families washing and playing in the water. About mid-morning, my stomach flared up. I felt incredibly guilty for running into a small roadside restaurant and essentially ruining their toilet, but it was the only option. As a Brit, I may have had a different relationship with the bathroom than people in Central America. We may view it as a sanctuary to deal with the unspoken, here is it just a place you go to do your business.

I managed to eke out 30km of running before reaching a small junction with no energy and a desire to just collapse. The only place I could find to sleep was one of the hotels that I feared also doubled up as a brothel but in my weakened state I just accepted whatever I could get. My fears were realised later that evening when a group of about four people arrived. Initially they sat drinking and laughing in the room next door but then headed to the bedrooms. It was immediately obvious that the walls were paper thin, so I quickly reached for my headphones. Fortunately, my body was so exhausted that even fornicating locals didn't prevent me from falling asleep.

Refreshed, despite my seedy accommodation the previous night, I started to appreciate that the El Salvadorian scenery was seriously impressive. Distant volcanoes came into view and a new sense of adventure erupted in me. Yet the last few days had been wearing and I didn't want to push my luck, so called it a day at the town of Zacatecoluca, about 25km south.

During my time on the road, I had learnt that a wake of vultures on the side of the road normally augured something gruesome ahead and today's scene was particularly disturbing. The large, black scavenger birds seemed to be collecting around a bridge and I peered over to find hundreds of bags full of animal carcasses.

On closer inspection, they were horses with hooves protruding, the bones stripped of all meat and emitting a putrid stench. A little online research revealed that the El Salvadorian gang, MS-13, had been selling huge amounts of horse meat as it tried to diversify its income into semi-legal and legal activities. Presumably, these carcasses were the result of this trafficking and this was essentially a mass grave. It was very harrowing to witness, especially to someone who has been surrounded by horses all his life.

Zacatecoluca lifted my spirits as it seemed to be in fiesta mode and I soon realised that it was Good Friday, a big event in a Catholic country. After finding a decent place to stay I headed into town to restock, refuel and explore. Preparations for the night's celebrations were underway and the streets were being decorated with intricate designs made of coloured sand. The images ranged from religious art to Che Guevara and even advertisements for local firms, focused around a set route where a procession would march. That evening I returned to join in and was bowled over at the sheer scale of the endeavour. This seemed to be a must-attend family event and every corner of the square was heaving with people. Families snacked on churros, grilled maize, ice cream and cake washed down with coffee or sodas but no alcohol.

The main procession followed a glass case containing a statue of Jesus Christ, surrounded by candles and held aloft by a large group of men in white robes. To avoid disaster, other men went ahead with wooden poles designed to raise the power cables and thus allow the procession through. The glass case was followed by statues of Mary and other important Catholic figures, and the whole thing wound its way around the town and then into the main cathedral. One family kindly asked me to watch it with them and I marvelled at everything that was going on. As the only tourist here it was a huge honour to be able to witness such a procession.

On spotting a news team recording the event, I seized the moment by walking straight up to the lead reporter and introducing myself, explaining my presence there. They jumped at the chance to interview me and seconds later I was being interviewed for the local TV channel. This proved a real test for my Spanish! My reasoning was to get as much recognition as possible, as the more visible I was, the more people might react positively to me, and in turn, the safer I'd be.

After the night's celebrations and a good night's sleep, I was ready to hit the road again and was feeling stronger than I had for days. The road seemed easier despite its ups and downs as the spectacular rivers, volcanoes and sprawling emerald-green fields inspired me to push on. Strangely, amid such boundless beauty, the sides of the road were strewn with litter. People seemed to have no issue with chucking rubbish everywhere and it was common to see passing cars discarding their waste into the fences and shrubs.

After 57km of running, I finally arrived in Usulután to have to face the usual logistical dilemmas of where to sleep and what to eat and drink. My research before arriving made me think that things would be easier here but I wandered aimlessly around, before settling for a small place on the outskirts of town and had to make do with the limited food sold by a local petrol station. An otherwise mundane stay in this town shifted gear when, in the middle of the night there was a sudden bang and my room started to shake. At first I thought it was a gunshot but later found out that it had been normal seismic activity from the local Volcán de Usulután. This was my first taste of volcanic drama but would not be the last.

San Miguel was my last town in El Salvador and I decided to reward myself with a day off before making the final push to Honduras. I hate to admit it but I was embarrassingly excited by the presence of recognised fast-food restaurants, which back in London I might have spurned. I was only in the town for 36 hours but still managed to fit in a visit to Burger King, Pizza Express and KFC. It wasn't a craving for this kind of food that drew me in but more the opportunity to be able to eat food that otherwise hadn't been available to me. Junk food aside, San Miguel was a charming provincial town and its main square was once again full of people. Here I could easily have spent the day people-watching but as always I had admin to get on with. In particular, there was needing to work out the postal system to send postcards home and finding spare parts to keep my 'well-used' stroller in working order.

The road always throws up some interesting and quirky things and en route to the Honduran border, it was a house made entirely of plastic bottles. Jewel-bright, the walls and roof of this house were made of green bottles all tied vertically together and inflected with little white spots of paint that gave it a gorgeous pattern. It was both ingenious and beautiful. A path had been created by decorating cement with different coloured bottle tops. Inside the house was an old lady sitting on her hammock, with a proud smile across her face. In stark contrast to the degradation of the countryside, here was someone taking the rubbish from the roads and using it to create a functioning building.

With the border to Honduras over 60km further south I hadn't expected to make it in one day but when I set off from San Miguel I felt refreshed and eager to be running after my day off. Rather than hit the tourist spots of a coastal road, I opted for the northern route, secretly hoping it would throw some extra challenges at me. Perhaps it was a blessing that it was in fact relatively smooth sailing. Santa Rosa de Lima was about 40km away, so it was perfect for a day's running. When I reached the Ruta Militar, I was hit by that familiar wave of excitement of being back on the road. I'd come to love what I did every day, even to the point when days off felt nothing more than a waste of time. When I was running I was alive, like being addicted to a mild drug. The realisation of not knowing what lay ahead was so exhilarating that I just wanted more.

Behind me, Volcan San Miguel slowly diminished in size as the day ticked by. The trees on either side of the road leant sociably inwards, creating an almost tunnel-like feeling in places and providing much-needed shelter from the oppressive sun above.

Time seemed to lose all meaning and when I looked at my watch at Santa Rosa I was surprised that I had got there by 2pm even though my legs felt as fresh as ever. I popped into a Subway to pick up a sandwich and decide my next move. The border was only 20km away and while my map said there were no hotels I was unconvinced. The challenge of getting to the border that day was irresistible, despite not knowing if any accommodation would be available when I got there.

In the final straight, local people assured me that there was a hotel called The Two Brothers. They talked of air conditioning and even introduced me to the owners' young son. Relieved that accommodation was sorted, I kicked on to the border and finally arrived at the hotel at about 7pm. What should have been a joyous moment was replaced with disappointment. The hotel was closed and appeared it had been so for some time. I guess I had been the subject of some kind of prank that had left me standing at a hotel on the border with nowhere to stay. There was only one option—cross into Honduras and find out what was on offer on the other side. This threw me as I had developed a routine before entering a new country and had not even thought about the alternatives. In the absence of practical solutions, I leant on spiritual means, praying as I crossed the border that the local currency was dollars. Of course, it transpired this was not the case.

The local moneychangers saw me as an isolated lamb ready for the kill and started to circle. Starting strong, I picked one guy and sent the others away saying I would only deal with him. He told me the exchange rate was 16 lempiras to the dollar. Much to his annoyance I pulled out my phone and with the tap of a few buttons found out that it was actually more like 21. Less confident now, he smiled and said he would trade at 20. I decided to take a stand and pointed out that he had just essentially tried to rob me and there was no way I was going to trust him. Irritated, I turned my back on him, sought out another and struck a deal at 21.

The border crossing was surprisingly easy and within five minutes I was standing in a new country, desperately trying to work everything out. Initial impressions confirmed the sad truth that Honduras was an even poorer country than its also poor neighbour. This was no time for sight-seeing, however, and a building on my right as I crossed the border hosted a hotel, supermarket and restaurant. I decided that after 57km I would defer the excitement of discovering Honduras until tomorrow and slept like a log in the sixth country of my trip.

CHAPTER 14 - FIVE DAYS IN HONDURAS

DISTANCE TO BUENOS AIRES: 9,591KM

Most people seem to avoid the short stretch of Honduras that sits between El Salvador and Nicaragua and the cyclists and motorcyclists who do pass this way write blog posts that suggest crossing the country as quickly as possible. Being on foot, I didn't have that luxury and I'm glad I didn't. At best, I could have done it in three days but decided to take it slightly easier and spread the journey over the next four days while trying to stay in all the main towns. Despite not liking towns, there was a particular rationale for this, because it increased the opportunities of finding secure accommodation and reduced the chance of being left vulnerable. In 2013, Honduras was named one of the most dangerous countries in the world with nearly 90 murders per 100,000 people, five times that of Chicago. The main reason was drug trafficking and cartel activity, which might explain most tourists' reluctance to stick around for too long. At the time of writing, different pressures mean that other countries are now ranked far more highly on this index than Honduras, including Mexico, which in 2015, was relatively trouble-free.

Although Honduras declared independence in 1838, it is the second-largest country in Central America and is split into three parts; Caribbean Honduras, Honduran Highlands and Pacific Honduras. I planned to run exclusively along the Pacific Coast.

Despite sharing a border, the scenery was noticeably different from El Salvador. Under the relentless sun, the land was dried to yellowish-brown, dead grass and leafless trees and shrubs dotted the hills, and pyramid-like hills pointed to the sky on either side of the road. Simple houses all faced the road, from which families spilled out, either relaxing in hammocks or going about their daily chores. Everyone was very friendly and there were lots of waves, whistles, shouts and smiles as I passed. Here too, the roadside was strewn with rubbish of all kinds. Plastic bags hung from tree branches and bottles lined the drains. It was heart-wrenching to see such beautiful countryside so horrendously abused. Wherever I went I saw litter flying out of passing cars and buses and people just dropping their garbage at their feet at the roadside. Signs asking people not to throw rubbish were ignored.

My body started to give up at about the 25km mark and the last 8km to Nacaome was a struggle. When I eventually got there, I found a decent-sized town that hugged a low hill to the south of the highway.

The town itself seemed split into two very different zones. The first was the market, where stalls lined the streets and sold anything and everything. Fruit and vegetables predominated, but mobile phone shops were a large presence, as were fried chicken restaurants. Further into town, things became more picturesque with a large white church presiding over the central square, and pedestrianised avenues of trees lined with benches. As with many of the towns I passed through, there was a sense that Nacaome had seen better days. The buildings looked tired and, in some places, had given in to the ravages of time. That said, a pleasant charm still permeated the place and the people were kind, helpful and as always, smiling!

My most pressing problem was that my stomach issues just wouldn't go away. There were two ways to deal with it: either lock myself away for a couple of days and recover, or just get on with things and suffer any potential consequences. Those who know me will not be surprised that I chose the second option. Eating was essential to keep up my energy, so despite feeling particularly unwell, rather than cooking my own dinner in my room I set out to experience some Honduran food.

Dismissing the restaurants as all pretty basic and on the face of it not that inviting, the lure of street food on the main square was hard to resist. This might not have been the most sensible option for someone with a delicate stomach, although it sure felt more adventurous. After a lap of the square to determine what was on offer, I settled on a restaurant called Pamela—a pop-up stall on a corner on the basis of it being the best lit, having the most customers, and that the staff wore matching uniforms—all good signs! The long cooking area was split into two stations. On the left was a hot plate with sizzling, marinated pork and gringas (tortillas filled with a tomato-based sauce, cheese and then fried), while on the right was the deep fat frying section, where I could see plantain and French fries cooking with a kind of chorizo. I opted for the pork and a huge portion arrived in a large polystyrene foam box, also containing plantains, which were fried and sliced length-ways, and piled high alongside the coleslaw on one side and succulent meat on the other. A combination of mayonnaise, tomato ketchup and barbecue sauce had been liberally dribbled on top. It was an amazing meal and even if risky, I figured it was well worth the consequences.

Amazingly, my stomach settled overnight and the day ahead would be brief, as well as having the bonus of being relatively flat. My target was the seaside town of San Lorenzo. My imagination painted beautiful beaches and beach huts in my mind, but the reality was somewhat different. It was lunchtime when I arrived, so the best option was the waterfront where locals had told me all the restaurants were. Sadly, it was evident that the tourism industry had been in decline for some time, and grand hotels that had once flourished were now run-down and falling into disrepair. Instead of the five-star experience that would have greeted travellers in the past, after a good lunch I had to settle for a grubby place on the highway, miles from anywhere. That night, I ventured back to the waterfront for dinner and was pleasantly surprised at how it had come to life in the moonlight.

Choluteca was to be my last main town in Honduras and there was no way of knowing what to expect. The run there was uneventful, apart from the moment a very kind fruit stand owner gave me a free banana on arrival at the outskirts. Once again, the major hill to climb was finding somewhere to sleep and as always, the bigger the town, the bigger the problem. To add to the fun, extracting cash from ATMs was also proving difficult. No cashpoint accepted my card and, stroller in tow, I walked from one part of town to the next in search of a machine that would work. The silver lining was that I had to navigate the streets through different areas of the town, and in the process stumbled across a restaurant serving the most succulent chicken I have ever tasted. The clientele was incredibly eclectic, bringing an odd mix of older ladies and biker gangs together over a love of eating.

Fortunately, I found a comfortable hotel on the outskirts of town that even had a swimming pool. Although more expensive than I intended to pay, the hassle of trying to find somewhere in this sprawling town had worn me down and the fact that it accepted card payments swayed me. Sometimes pragmatism must rule the day and it is just worth paying more to have more time to relax and recover.

The hotel owner tipped me off about the presence of a large mall in town. The first impression certainly did not disappoint. This place rivalled London's Westfield centre in size and cleanliness, and it was full of shops I certainly couldn't afford. Locals sat and chatted, enjoying the air conditioning, American brands and fast-food stalls in an utterly different world to everything I had seen in Honduras during my short stay. There was no doubt that Honduran society contained layers I would never understand, and that inward investment was making places like this possible despite the decline I had experienced elsewhere. The mall briefly invoked a kind of nostalgia and I gave in to the sense of familiarity for a few short hours.

My short time in Honduras was already coming to an end. It felt good to have proved the naysayers at least partially wrong about this country. Everyone was friendly, smiled and waved, and even Carlos the bear got a lot of attention from ladies and children alike. The road today wasn't too bad as it had a hard shoulder of sorts for me to run on, and green trees either side of the road provided shade. On the way south I tried a new fruit, a pupunha or peach palm, that

I bought from a stall on the side of the road. It was red and looked a little like a red pepper but inside the flesh was sponge-like and held a lot of liquid. Apparently, it's rich in vitamins A and C, selenium and nicotinic acid, as well as protein and starch—all good for a runner's diet.

On my approach to the border, I came across a bunch of lads trying to sell large, live iguanas to passing travellers. To prevent them escaping, the iguanas' legs had been bound behind their backs, and I was tempted to buy one just to release it but realised how ineffective and provocative that would be. Culturally, I was a visitor only with no right to interfere.

At the border, dreary lines of lorries waited to cross, while bars served their drivers and men touted currency to travellers. I rehydrated with a cool beer while I got my dollars converted into Córdoba, the local Nicaraguan currency. The border crossing itself was a relative breeze, except for the time my immigration officer inexplicably went walkabout halfway through the process.

After negotiating border control, I had a nice easy 6km run into the small town of Somotillo. Despite being set a little away from the actual border, it provided the essential services. If I thought things had become a little more basic going from El Salvador into Honduras, the same thing happened on my arrival in Nicaragua. My hotel, the Fronteras, was very rustic but once again oozed a kind of splendour that had long since been lost. I loved it.

Preparing to settle in, however, produced a nasty surprise. At the end of every day, on reaching my destination, I would check in with my GPS tracker and tonight was no different. Except that when I went to look for it, it wasn't where it should have been. In the absence of lithium AAA batteries, I'd only been able to check in at the end of, rather than throughout the day. Each night, therefore, I needed to find somewhere with a signal to check in properly. This tracker was primarily a safety measure and it allowed me to send a standard email to my father to let him know I was safe. There was also an emergency button to press should something go wrong, though I'm dubious how quick the response would have been.

Losing it was a serious annoyance. I rummaged through my stuff in an attempt to find it and then remembered that during my last night in Honduras I had placed it behind a curtain (which was the only place it could get reception) and maybe I hadn't retrieved it when I went to leave. Thoughts ran through my mind. I needed the GPS to continue and could either contact the hotel and ask them to put it on a bus, or get on a bus myself and return to Honduras to collect it. On balance, there was only one viable solution, so I wearily prepared to return to Honduras.

CHAPTER 15 - BLOOD, PEE AND VOLCANOES

DISTANCE TO BUENOS AIRES: 9,451KM

Boarding a bus back to Choluteca was not how I envisaged starting my time in Nicaragua, but it was undeniably essential. Having successfully navigated all the border procedures the day before, doing it in reverse was relatively straightforward. There was no way this detour was going to be allowed to affect my overall mood so instead I embraced the journey as an opportunity to be a tourist. Like all the buses in Nicaragua and Honduras, this one was a decommissioned US school bus that had been given a new lease of life by replacing the distinctive yellow with bright colours, names and logos. These buses are lovingly referred to as 'chicken buses' and are packed with people, bags of vegetables and whatever else can be crammed inside, including live chickens. Every now and then someone would make a sales presentation to the passengers and then pass through the bus handing out pens, sweets or whatever else they were trying to offload.

After a successful mission across the border and back, my focus reverted to reclaiming the road. I was 70km north of a town called Chinandega and set my sights firmly on getting there the next day because I had heard from people on Facebook, notably Brooks the cyclist I'd met in both the US and Mexico, that a Spaniard called Nacho was walking his way around the world and if I could reach the town the next day we would be able to meet. Brooks assured me that we would get on and encouraged me to make the effort.

In the soaring heat, the only way to lessen running in the midday sun was to rise early and get on the road. The route wove between the Reserva Natural de Apacunca and the Reserva Natural Complejo Volcánico Cristóbal Casita, home to the highest volcanoes in Nicaragua, and also some of the most active. Running through the parched countryside as the sun gently rose was breathtaking. With my speakers on full volume, in my upbeat mood as I sang along to whatever played next. At one moment I was so engrossed in a full karaoke rendition of Enrique Iglesias' Hero, that I was oblivious to a camping-car drawing up alongside with a young French couple staring at me in bewilderment.

Water was, as ever, the biggest concern but luckily small stalls were pitched along the side of the road wherever shade could be found, and people sold water in small plastic bags rather than bottles. Sadly, it was these small blue bags that littered the sides of the road and I can't help but think there must be a better alternative.

After about 20km it was possible to make out a volcano in the distance peeking through the mist below. This perfectly conical volcano became my focal point that day and as I marvelled at how imposing it was, wisps of steam rising from the crater continually reminded me that it was active. The road climbed and descended over small ridges, and as it did the conditions would change from parched and arid to lush, green and agricultural where horses roamed freely. People could be seen working the land or resting on the porches of the wooden homes that looked over the road. At one point, a gentleman on an old rickety bike slowed down just behind me, and without uttering a word followed me for about 8km.

It was at about the 40km mark that I started to realise that something wasn't quite right. In a vain attempt to stay hydrated I'd been drinking a lot of water because I was sweating profusely, streaking my shorts with white salt marks. Despite an unquenchable thirst, I found I needed to pee and stopped behind a tree on the side of the road. Looking down, my urine had turned bright red but I dismissed this as in some way connected to the Gatorade I had just drunk. Another kilometre or so further down the road I suddenly felt a desperate urge to pee again but this time only a red dribble came out, accompanied by an excruciating burning sensation. I should have sought a solution there and then, but being in the middle of nowhere and with a single-minded determination to get to Chinandega, I continued on. With nearly 30km still to run I had to dig deep to find the strength, which was only made harder by an uncontrollable urge to pee every couple of kilometres. Somehow, I made it. When I got there the last thing I wanted was the ordeal of finding somewhere appropriate to sleep, so just opted for the first hotel and set about taking on as much liquid as possible in a vain attempt to stop whatever was wrong with my bladder. To hydrate, I bought large, cold bottles of water, Coca-Cola to kill any germs in my stomach and beer to relieve the pain. Any doctor reading this may question these choices, but the thing about being tested to the limit is that it becomes imperative to resolve issues as they arise and to do the best you can with the resources you have to hand. Despite all that was going on with my body I was still determined that I would make my dinner date with Nacho, so I sent an email and waited for his reply.

Nacho and I finally made contact and arranged to have a burger in a street market in the town centre. I had no preconceptions but felt instantly uplifted to meet someone who was experiencing the same challenges as me. We had similar stories, issues and feelings about our adventures. Nacho was a great guy with a positive outlook on life and his adventure and it was insightful to talk about how things affected us and to gain another perspective on how it felt to be alone on the road. Recalling this little encounter now, I can see that taking more time with Nacho would have been a good thing but at that moment I was so focused on moving forward that I was blinkered to the bigger picture. Here was I, continuously feeling out of place, lonely and suffering from a potentially serious bladder issue passing up the opportunity to spend time with someone who would really have understood my situation. I will never know exactly why, but when I look back, I wish I had made a different decision.

It was some relief to wake the next morning and see that my waters were relatively clear. Thinking I was out of the woods, I packed my stroller and hit the road. The sensible thing to have done would have been to rest for a day and see a doctor, but as established, I was blinkered. Not long out of the town I immediately regretted my decision, but rather than turning back and admitting my vulnerability, I forged on, despite an agonising need to pee continually. At the 5km mark I found a tree, where I was aghast that my urine was once again bright red and passing it caused the same burning pain. There was only one logical explanation as to what it was and that was blood. Deep down I knew but didn't want to acknowledge this, so staggered on trying to work out what to do despite feeling seriously ill and most probably running a fever. My options were to double back or attempt to kick on for 38km to León. This combination of determination, pride and stupidity precluded rational thought but luckily a sign by the side of the road for a pharmacy with consultancy services gifted me a third option. Some way down the dirt track was a little red bungalow displaying a pharmacy sign outside. Inside, a friendly lady introduced her husband, who turned out to be the doctor. I wasn't the only one seeking his attention and was asked to wait while he dealt with a chap who had a huge gash on his arm as a result of a fight with chainsaw.

Stepping into the doctor's office, I had a premonition that this was going to test my communication skills. In broken Spanish, I tried to explain.

'Yesterday, while running, I experienced an urgency to go to the toilet. When I went it was incredibly painful with a sort of burning sensation and my urine was bright red. This continued about five times an hour with the amount of urine reducing to nothing more than a dribble.'

He shot me a quizzical look and it dawned on me that I may have been describing symptoms that sounded an awful lot like a sexually transmitted disease.

To clarify, I added, 'This can't be anything sexual, as I haven't had sex in months...' Things were awkward for a few seconds but then he picked up a small jar that looked very much like it had been used for jam and led me out to a rose bush at the bottom of his garden. 'Please make a urine sample into this and bring back to me.'

Feeling a little self-conscious, I did as instructed and took the sample back to his office. About 20 minutes later he returned with a piece of paper containing a full analysis of my urine. 'You have a urinary tract infection and cystitis,' he informed me, as he prescribed antibiotics. 'It should clear up if you take these pills, which my wife will sort out for you.'

In normal circumstances my question would have sounded crazy, but I asked anyway. 'I am trying to run to Panama. When will I be able to continue running today?' My apprehension dissolved when he replied, 'Take it easy for the rest of the day, rehydrate, take your medicine and you should be ok to continue tomorrow.'

Apparently, my infection was caused by a lack of cleanliness down below. Normally, ladies are more prone to infections like this because of the proximity of the anus and the vagina, while men are protected somewhat by the greater distance between the anus and the head of the penis. In my case, the combination of a lot of sweating, not enough showering and lycra shorts meant I was at increased risk. I hadn't even considered that my lack of hygiene would lead to me actually getting ill and this is a classic case of learning on the job. Fortunately, there was a hotel next to the surgery and I spent the rest of the day following the doctor's orders.

Trying to put the discomfort of the last couple of days behind me, I focused on getting some kilometres on the clock. Thankfully, my arrival in León a little after lunchtime certainly helped take my mind off recent woes. León's architecture is mainly colonial with some amazing buildings making up the city centre, including its cathedral, which is a UNESCO World Heritage Site. These ancient edifices blended fantastically with the chaos of the new.

Many things attract travellers to León, including the food, the party atmosphere and a rest from the beach, but it is also the home of a crazy sport known as volcano boarding that was rated number two on CNN's list of the most electrifying things to do—sliding down an active volcano on a piece of plywood! Of course, I signed up to take part next day.

That afternoon I bumped into Ellen, the Scottish girl I had met at the border into El Salvador, and we decided to explore together. First was the Centro de Arte Fundación Ortiz Gurdián, widely accepted as one of the most important museums in Central America. Once we'd had our cultural fix, we made our way to the Cathédrale de León for a quick tour. For a few dollars extra, we were allowed to climb up onto the roof of this amazing building that occupied pride of place in the centre of the town.

Playing the tourist was a helpful distraction from my usual onslaught of routine and running and I freely acknowledge that I probably wouldn't have done half of these things if it hadn't been for Ellen's itinerary. Our day ended over a couple of beers watching life passing by in the town square while eating dodgy street food. I wouldn't have had it any other way.

The morning began full of trepidation about the day's activity, made worse by a fogginess induced by a few late-night drinks in the hostel the night before. Nonetheless, I joined the group piling into a battered pick-up and we headed out of town towards an active volcano called Cerro Negro. On arrival we were each handed a long wooden board and an orange jumpsuit, similar to those in American movie prison scenes. Cerro Negro loomed above, daring us to enter. The 700m climb wound up the black volcanic rock and as we crested the final section, looking down on the steam vent truly felt like a different planet. The wind wafted the putrid smell of sulphur around and heat literally rose up through the ground, warming us from the feet up. This environment was harsh, and here I glimpsed the raw power of nature.

Let's be honest, volcano boarding doesn't require much skill. From the top of the slope you can see the route straight to the flatter section below. The aim is to get from the top to the bottom as quickly as possible without being separated from the board. One of the guides stands halfway down the slope with a speed gun to measure boarders' speeds. One by one, we positioned our boards on the lip of the slope, sat down, pulled on the protective glasses and nudged ourselves forward and over. It takes a few shunts to get going but then the speed quickly builds and soon adrenalin is coursing through your veins as small rocks fly up and you hurtle down. The record is about 100km per hour, but I am confident I got nowhere near that. The rush was immense and the perfect way to break the routine of daily running.

Back at the hostel, still on a high, a small party started to form and of course descended into the usual backpackers' afternoon drinking. One of the participants was a British guy who worked with a few brands and had connections to the media back in the UK. He took a few photos, made some notes about my journey and a few weeks later managed to engineer an article in the Daily Mail. And it didn't end there, as before I departed the next morning an American photographer called Cris-Ian got up early to take a series of images that illustrated the many articles that were published once I had finished.

Hanging around people who all seemed so connected, combined with drinking too much always seemed to lead me to a melancholy place, as it had even before this adventure began. Seeing people happy and with their partners reminded me of just how lonely my life on the road was. When sober it didn't bother me, but late at night after a few too many drinks it always left me feeling low. On my final night in León, I spent a couple of hours lying in a hammock alone, reflecting. Taking advantage of the Wi-Fi, I called my ex-girlfriend and spoke to her for hours, which, while cathartic, just made me feel less connected to anything. My journey had taught me how to deal with these moments and the solution was simple, get back on the road and let the adventure provide the distractions, which I guess just highlights how much I avoided the issue, rather than confronting it. It begs the question, was this very journey my way of avoiding my troubles and keeping my demons distracted?

These couple of days in León completed the rollercoaster of emotions that comes from going from solitude to being around people. I craved the closeness, enjoyed it, bathed in it until it dragged me down, before feeling the familiar urge to get away. With no particular plan for the day ahead, I just knew I had to get back on the road and re-discover the local culture. That night, my penance was to stay in a small hotelito on the back streets of La Paz Centro, a basic little town just off the main road with little more than a service station to provide rations for the night.

Managua had not had great reviews from other travellers, but it was on my route and I decided to spend at least one night in Nicaragua's capital city. The distance was just under 60km and the route took me along the edge of Lago Xolotlán but I didn't get to see much of its beauty apart from the ever-present volcanoes in the distance.

As I got closer to the bustling capital, the little settlements got progressively more urban and the roads busier. In general, running into big cities can be stressful with all the cars speeding by as I am normally at the end of a very long day of running. Here that wasn't the case. The town slowly built up around me and I used the ever-changing surroundings to discharge my usual tension. On the outskirts, the foundations of new housing projects were being laid everywhere, but strangely right next to schemes that had obviously run out of financing. My route took me down to the docks and then through some residential parts of town, which gradually got smarter the closer I got to the more affluent part of town. Despite this city clearly being quite a lot poorer than the others I had entered, I never felt threatened.

Once settled into my hostel, my GPS tracker produced another drama. As always, I had placed the device out in the open to check-in before taking advantage of the swimming pool after a long day on the humid road to Managua. When I went back to retrieve the tracker, it had gone. Being overtired and probably a little irrational, I slightly overreacted and started a frantic search of the hostel. Ironically, the only thing I seemed to repeatedly lose was a device that was designed to track my every step!

One of the guys by the pool noticed the security cameras so I went in search of the hostel manager. The other guests joined in the hunt and one indicated that he had seen a large drunk American pick it up and inspect it. Unfortunately, our main suspect was passed out

on the sofa, so I had to wait for him to wake up before confronting him. 'Have you seen a small orange GPS tracker?'
'Never seen it, don't know what you are talking about,' he snapped back.
'Are you sure? We have trawled through all the security and there are some pretty clear images of you picking it up?' Another flat denial. The search continued while we continued to press for any admission of guilt. Although just a small tracker, to me it was a lifeline and when I decide to recover something I am relentless in that pursuit. 'If you just tell me where it is, then we can forget anything happened!' Nothing.

Finally, the tracker is located in a cupboard in the guest kitchen. To be honest, I don't think the man did it maliciously but was just too drunk to remember what he'd done. Unfortunately for him, the management decided that his actions and general drunkenness were not going to be tolerated any longer in that hostel and he was politely asked to leave.

My default when it comes to planning my route is how to get from A to B by the most direct means, but on occasion, there is something worth making a little detour to explore and the old port town of Granada was one of those exceptions. This beautiful colonial town had witnessed much action over the centuries and was attacked by the English, Dutch and French before being burned almost to the ground by filibuster Charles Henningsen in 1856 as he retreated. Very little was left standing and as he boarded his boat, he left a lance with an inscription saying, 'Here was Granada.'

My run to Granada featured a strange little run-in with a Guatemalan. Passing through the town of Masaya, I got what I thought was verbal abuse from a young man who seemed drunk. As usual, I just ignored him and ran on. Later that afternoon a bus drew up alongside and the same guy got off and came over to me, declaring that he wanted to run with me despite being dressed in jeans and flip-flops. Not wanting any sort of friction, I said he could join me if he could keep pace with me. As it turned out, he was an acrobat in a travelling circus that was currently based in Granada. Our conversation ranged from life in Central America and what it was like to be part of a travelling circus to life in London and just how different it was. The dynamic was odd, in that I was running away from the very place he could only dream of being in. Clearly very fit, he managed to keep up with me and after about 10km we arrived in the centre of town. Parting company was awkward as he clearly had nowhere to go and very little money. Despite feeling incredibly guilty not to have given him anything, I had to be resolute in not giving handouts.

Granada was a beautiful town so I plumped for another day's rest and exploration despite a creeping guilt over the quantity of rest days I'd taken in the last couple of weeks. With hindsight, that seems far-fetched because I had run an average of 42km a day for 11 of the last 13 days, clocked up over 55km on three occasions, taken two full rest days and had contracted and overcome a urinary tract infection in the process. Luckily, the beautiful churches, hectic streets and bustling markets were enough to distract me from beating myself up too much.

The border with Costa Rica, Nicaragua's more prosperous neighbour, lay just over 100km to the south, but I was enjoying Nicaragua so much I was in no rush to leave. Moreover, a little treat was in store before I crossed the border.

On the way out of Granada, I had my first stroller race of the adventure. While ascending a gentle hill I started making ground on a local chap who was pushing a two-wheeled wooden cart. His cart was full of water containers, topped with a lady about the same age as him, who I assumed was his girlfriend. As I drew up alongside, we locked eyes and smiled because the same idea was running through our minds. The pace quickened. Passers-by smiled as this rather odd-looking race played out. Foolishly, I thought I had a clear advantage but was soon proved wrong as he pulled up alongside. We veered towards each other, jubilantly high fived, abandoned our race and then set about our respective days. Small interactions like this make journeys special and even better. A little further down the road a father and his son stood on the side of the road with a bucket of cold water ready for me. They had spotted me earlier and had set up a little rest spot for me to refill my bottles. A small act of kindness but one that I'll never forget.

The road out of Granada continued to throw up amazing little glimpses into Nicaraguan life. Huge white oxen pulled carts laden with harvested sugarcane down the gravel streets while old ladies drew water from wells in the most ingenious way. In groups of two, the first would tie one end of a rope around her waist and the other end to a bucket that was dropped down the well. She would then walk away from the well, hoisting the bucket using her body weight as ballast up to her companion waiting at the top who would then pour the water into another container, before they repeated the process.

By mid-afternoon I found myself in the village of Nandaime and asked around about places to sleep. My choice was a hotel that was just 10km out of town and on my way to Costa Rica. When I got there I realised that it was somewhat smarter than I was used to and therefore out of my price range, but the sight of a beautiful swimming pool swayed me to fork out the extra few dollars. On being shown to my room, the manager came rushing over shouting that my stroller was not allowed in the bedroom. By this point, I had developed a rather strange relationship with my stroller and would not be separated from it during the night unless absolutely essential. The Hilton and the Camino Real had let me wheel it into my room so I was damned if this little hotel on the side of a road in Nicaragua was going to refuse me. Moments later I was in the reception demanding my $25 back.

This rash decision presented me with a slight dilemma. It was mid-afternoon, I had already covered about 30km and Rivas was nearly another 40km away. There had been murmurs about a place about 10km down the road, so I pushed my stroller back into the searing heat and started to run again. When you have convinced yourself that you have finished for the day, mustering the motivation to run another marathon distance can be hard, and especially so in the heat of the day. I managed, but once again was disappointed when there was no available accommodation. My saving grace was a chicken burger and to refuel for the remaining 30km stretch to Rivas.

When the world seems to go against me, I invariably find a new sense of determination. I like the challenge. Not because I want to shout about it but because it's a hurdle that needs to be overcome. My mind becomes preoccupied with working out how to tackle whatever it is and then I just close down and focus. On this occasion, the solution was to break the run into three segments and limit myself to 10km per hour. It helps to find things to take my mind off the pain and on this stretch of road, it was a horse-drawn cart with two young guys on the trailer. I would catch up with them and try to overtake before needing to take a rest, they would then retake the lead and give me something to focus on. This simple game played an important part in eventually making it to Rivas.

Rivas is a small town that sits on a crossroad between roads to the Pacific Coast and the vast Lago Cocibolca, as well as the main route to Costa Rica on the Pan-American Highway. By now it was dusk and there was little more to see than a few rundown hostels, restaurant shacks and service stations. After an epic day like this one, securing my stroller became the priority before ambling around the streets and enjoying street food and local beers. This time is for reflecting on just what is possible when you find yourself with very few options. Running big distances can be hugely enjoyable, but they are even more rewarding when unplanned.

Costa Rica was within easy reach, but I had options to extend my time in Nicaragua. To the west a few surf towns perched on the Pacific Coast, which were highly rated by other travellers I had met. To the east was Ometepe, an island in Lago Cocibolca that was home to Volcán Concepción. Ometepe is renowned as a place for relaxing in hammocks and lapping up the laid-back Nicaraguan lifestyle. To the south was San Juan del Sur, a party town that was famed for its Sunday drinking sessions.

As luck would have it, a traveller in León had emailed the owners of a hotel and told them of my adventures so I was invited to stay for free with them in the hills overlooking San Juan and the ocean beyond. Thus it was decided, I would let my hair down one last time in Nicaragua.

My hosts were the friendliest people imaginable. They immediately took me under their wing and showed me to a dormitory and invited me to stay for the next couple of nights. The main dorms edged an infinity pool that boasted the most amazing views, especially as the sun set over the ocean below. There was a real family vibe, and everyone seemed completely at home. Rather than providing a cafe, in the evenings anyone who wanted to eat could attend what was termed the family meal around a single table. Inevitably, this led to drinking into the small hours.

Despite the hospitality and community feel, most of my time in San Juan was spent exploring. I wandered the streets and allowed its quaint atmosphere to soak in. This chilled surf town was tailored both to backpackers and more affluent holidaymakers, meaning that the small brightly-coloured wooden stores offered crepes, iced coffee, surf kit and smoothies instead of the usual basic provisions. The curved bay glittered under the protective gaze of a huge statue of Jesus, who peered down from a southern headland. Along the beach, bright fishing boats lay in a jolt of colour against the dark volcanic sand. It was the kind of place most people could unwind and relax, but for me, that didn't seem possible. The call of the road wouldn't quieten and even between sips of ice-cold Toña Cerveza, I was planning my route into Costa Rica. That night I retired from the partying early and despite numerous disturbances from guests returning to their lockers to snort cocaine, I tried to rest up for the crossing into my eighth country.

CHAPTER 16 - DIVING INTO GENEROSITY

DISTANCE TO BUENOS AIRES: 9,045KM

Compared with Nicaragua, the Costa Rican scenery wasn't dramatically different, and parched fields still ran up hills and volcanoes in the distance. But what was most noticeable was the cleanliness of the roads and the surrounding countryside. Throughout Central America, I had been saddened by the lack of respect the locals had for the countryside. Here things seemed different, more developed, more organised and most importantly, the people appeared better educated.

One key difference that affected my running was that the general economy in Costa Rica is far more established. This was instantly obvious in the more modern housing and better roads, and that people were driving better-quality vehicles on them. Being in Central America had fostered my reliance on the small shops that lined the road but here these were fewer and further apart, as they had been in the United States. The further south I went, the warmer and more humid it became and consequently, the more dependent I was on roadside provisions. I was learning that the frequency of shops directly correlates with the economy of a country. The more developed the country, the fewer small shops are needed but in poorer countries where people have less access to refrigeration and cookers, they are more reliant on the small roadside shops and restaurants. Clearly, a change to the way I planned and executed my day-to-day routine was in order.

After just 20km in Costa Rica, I must have looked rather forlorn sheltering at a bus stop with very little water and struggling to move forward. The heat was intense and the sun oppressive. One technique that provided relief was discovering that if I cooled down my core then I could start running again. Even so, I was going to need more water to avoid suffering the same dehydration issues as I had experienced in Nicaragua. The fertile, green surroundings seemed cruelly to mock my lack of water.

My only sensible option was to push on to Liberia where I knew a hostel waited. Its relaxed and convivial atmosphere comforted me to the marrow of my bones, while the great group of guests provided some much-appreciated company. Here too, were people from the UK who brought a welcome rush of familiarity. Sometimes being so isolated means that you close down. But after some proper food, good Chilean wine and fresh conversation I was revived, in body, mind and soul.

Mornings were generally regimented and alone, and opening steps were spent mentally focusing on the distance I needed to cover that day. While it was lonely at times, other days were filled with interactions that reaffirmed my faith in humans. There was a French cyclist who was travelling north, and later some friends of my soon-to-be-host in San José, and later still, a mother with her two children pulled over and handed me two large bottles of frozen water. None of them could even partly imagine the impact their presence in my life had at those moments, and I have almost no way of telling them now. It was these seemingly insignificant daily encounters that filled me with the positive vibes I needed to dig deep and drive myself forward. Part of me relished these moments so much that I had tricked myself into believing that the only way to experience them was by being out there.

On either side of the motorway lay great plains of farmland. Picture-perfect trees with beautiful peach and purple flowers interspersed with palm trees against a backdrop of looming volcanoes on the horizon. Perhaps the monotony was getting to me as a signpost to a waterfall was enough for me to decide to venture from the asphalt. Leaving the highway led to an uneven dirt track, which was not easy to navigate given all the obstacles, the heat of the midday sun beating down on my back and without the benefit of the breeze that ventilates the roads. At the car park a security guard agreed to look after my stuff. All the additional hassle was worthwhile, as the waterfalls had a rare and cinematic quality on an epic scale. The water flowed evenly over the rocks above to cascade like a curtain into the sanctuary of pools below. Families paddled as I clambered over the rocks and between the trees to find somewhere more secluded. After my morning's exertions, swimming was so refreshing and cool as I entered the wall of water and let it massage the stress from my body.

That night I barely made it to the small town of Cañas and honestly couldn't have struggled any further in the heat and humidity. Although missing the rustic character of Nicaragua, I couldn't deny that the everyday generosity of the Costa Ricans was overwhelming and I was almost certain it would define my time here.

I had been on the road for nearly 10 months now and although my friends and family were lending support from afar, during that time I had not met up with people I'd known in my past life in London. Here in Costa Rica, I finally had the opportunity to re-connect with a friend. Richard and I had originally met through a very good friend and he was now living in San José with his fiancée (now wife), Raquel. San José, Costa Rica's capital city, was not exactly on my direct route south, but I decided to make some allowances on this occasion. Together, we hatched a plan that I would run to his fiancée's parents' country house, which was on my route, from whence he would collect me and drive me to the capital. There was no stopping me now, as I had made good time and was within two days of arriving and a day ahead of schedule. The prospect of being with a friend formed a magnetic force that pulled me along the road, with only a couple of 50km days between us.

One of the bonuses of being in a more developed country was the presence of Subway. In America, this had been a staple addition to my diet, and it was great to have a nice foot-long sub with a side of chocolate cookies to fuel the engine. Nothing before my departure had prepared me exactly for the yawning hunger that running very long distances every day could induce.

Conversely, restaurants here were more discerning about letting people just camp on their premises, which made finding places to stay trickier. I'd stop at restaurants only to be told not to camp there, and perhaps I could try to find shelter at a place just a bit further down the road? After many rejections and with the light starting to fade, I luckily stumbled on a restaurant next to the highway that served the large tourist buses. My haven would be a small area of tiled flooring at the top of a small tower-like structure next to the restaurant. That night I was continually awoken as bus after bus rolled in and weary travellers stretched their legs before their journeys continued, unaware of an exhausted runner in the watchtower overhead.

The following evening, I finally arrived at the café where Richard had suggested we meet and awaited further instructions. His fiancée's family house turned out to be somewhat grander than I had at first imagined, especially in comparison to everything I had experienced over the last few months. On being admitted through the private security post and down the long drive I was greeted by Junior, the friendly caretaker. The house was out of this world, both architecturally and in its decor, and I felt rather out of place in my smelly running gear. Richard immediately took me under his wing, intuitively understanding how restorative a shower would feel before a dip in a swimming pool and an ice-cold beer. This lifestyle was worlds away from the dusty roads and I am not ashamed to say I was rather relishing it!

Raquel's father was an incredibly interesting man who, on our journey back to the city, filled me in on the politics of Costa Rica and the difficulties the current government was facing. Despite being more developed, it seemed Costa Rica still had its problems.

Ever since losing my bank card in Guatemala, four countries earlier, I had been trying to engineer a way to get a new card. In the intervening months I had asked Richard if he had had any plans to be in the UK and unfortunately he didn't but he would be seeing his brother Nick, who was UK-based, while in New York for work. This prompted a huge logistical operation that involved ordering a new card, having it delivered to the couple living in my flat in London, who in turn passed it on to my sister-in-law. She organised for this to be couriered to Nick who worked in Canary Wharf and who travelled with it to New York before passing it on to Richard who then flew it to Costa Rica. A huge relief for me and an amazing team effort all around.

Richard and Raquel had to work during my stay so I was free to roam the city. On my return that evening my hosts treated me to a delicious home-cooked dinner. Raquel's younger sister Juliana joined us, whom I found particularly great company, and we all enjoyed an evening of laughter and happiness.

To be reunited with my stroller I had to return to where it was being kept and that meant taking a bus from San José. My kind hosts gave me a lift to the bus terminal, which turned into a mad rush through the traffic and crowds, and even so I arrived at the bus terminal only to find out that my bus had already left. Another couple were in the same predicament, and we decided to join forces and hire a taxi to see if we could catch the bus. Like some madcap comedy, we frantically piled into the nearest one and set off in pursuit. Our driver was in constant communication with his friends trying to locate the bus but we had finally to admit that we weren't going to pull this off. This left us with a dilemma: return to San José and find another bus or just group together and get the taxi to our respective destinations. After negotiations, we managed to come to an agreement and continued south in the taxi.

Junior, the caretaker, was on hand to help prepare me to get back on the road. Just as I was leaving he introduced his young daughter and son and when his daughter saw Carlos on my stroller, she ran inside, came back with one of her teddies and presented it to me. I am not going to lie, it felt horrible taking a teddy from a little girl but she insisted. Her bear stayed with me for the rest of my expedition and now lives with me in the South of France.

Being away from the coast for this long meant I was missing the reassuring presence of the Pacific Ocean on my right side. The road met the coast at Jaco, a place renowned for being where

American ex-pats and holidaymakers get their weekly casino fix. To get there, I needed to take the bridge over the Rio Tarcoles, with hundreds of American Crocodiles lazing on the muddy banks below, unconcerned by the tourists that lined the bridge or the locals hawking trinkets.

As the road neared the ocean, the scenery changed dramatically from arid to greener jungle and the humidity rose to match. Macaws perched high in the trees above, screeching to each other as I ran by. Along this road was a Spanish cyclist called David who was on the first day of a cycling adventure that would also take him to South America. As he slowed down and cycled along for a while, he asked about my travels and outlined everything he hoped his journey would deliver. While we were travelling together a pick-up truck pulled over and a friendly young guy jumped out to offer us a cold drink each. We all got talking and unearthed an immediate mutual appreciation of adventure.

Our new friend put forward an idea. 'When you get to Jaco, I want you to go to a specific restaurant and say that I sent you. I will go there now and tell them and I will take care of everything. Once you have finished there, make your way to my friend's hotel and there will be a room reserved for you both. I love what you are doing, keep it up.'

At that, our benefactor got back into his pick-up, rendering us speechless as he drove away. We had a target for the day and a new motivation to get there. True to his word, when we got to the restaurant we found that our friend had been in and pre-paid for everything we could eat and drink. After we had eaten, at the hotel we were once again amazed to find everything had been taken care of. This astounding act of generosity came without conditions, and like all the others that made my journey more comfortable, was utterly humbling in its selflessness.

Over dinner I got to know David a bit better but he was suspiciously cagey about his past. It seemed like he had served in the Spanish Army, maybe something similar to the Special Forces. He was vague about details, so I didn't pry. All that mattered was that we shared a motivation and were using travel and adventure to find something more fulfilling or just understand ourselves a little better.

Our benefactor had published our journey on his Facebook page and we started to receive messages from others in the local area. One came from a restaurant owner in Parrita, about 40km away, and David and I decided to take him up on his offer of a free meal. Due to our differing speeds, I set off early and agreed to meet David there. Running out of Jaco was relaxing, enhanced by the calmness of the ocean, something I'd missed more than I'd realised.

By the time I arrived in Parrita, David had already embedded himself with the locals. He was one of those characters that could easily muscle into any group and become the centre of attention. We met the restaurant owner, who was incredibly hospitable and had seemed to have enlisted his entire extended family to look after us! Fresh juices made from fruits I hadn't even heard of flowed and the chef was keen for us to try everything on the menu. With full bellies and after customary selfies, we rolled into town and found a small hotel for the night.

David and I had spent the best part of two days together and over beers that evening decided to part company the following day. He was keen to start covering more miles and to be honest, I was used to being more self-sufficient. Having a travelling partner entails a lot of compromise and when you are used to being as single-minded as I was, you notice that.

Costa Rican hospitality seemed never-ending and a mere 30km further down the coast was yet another free meal. Raquel had spoken to a friend in a town called Quepos and arranged for me to have a meal in his amazing restaurant next to the marina.

With this relatively short day ahead I was able to take in more of the gorgeous countryside. On either side of the road were vast palm plantations where farmers worked with ox-drawn carts to collect palm nuts. I had very few touristy destinations planned along my route, as the whole point of my adventure was to keep on moving, although I did know there was something at Quepos I did want to explore, so I told myself sternly to go and do just that. My hostel was perched on the crest of a ridge overlooking the ocean, so enlivened by a delicious lunch, I battled one of the steepest hills I had yet encountered to get there.

The next morning I ventured 3km south to the Manuel Antonio National Park. It's so often the case that an attraction that is renowned for its beauty can create such a contrasting environment right next door. This was the case here, and Quepos had grown into a very touristy town in which the road was lined with tawdry resorts, supermarkets, souvenir shops and even a gentleman's club.

The huge diversity of wildlife is one of the main reasons people flock to this particular national park, which offers possible sightings of sloths, white-faced capuchin monkeys, howler monkeys, ghost crabs, racoons and iguanas to name but a few. Within seconds of entering the park, a white-tailed deer emerged from the woods and into my path, and I immediately got the feeling I was going to be lucky. I saw capuchin monkeys, a sleeping howler monkey, racoons and was awestruck by a huge iguana. Spending time in nature without the pressing need or desire to keep pushing on was pure catharsis. On little deserted beaches, armies of red crabs scuttled around scavenging for food and iguanas idled on black rocks in the blistering sun. The trees were immense, and the undergrowth constantly rustled with lizards and crabs on seemingly urgent errands. I used this special time to try to disconnect from the highway, the journey and the pressure I was putting on myself. Even if it was self-inflicted, a break from the pressure was still warranted.

Back at the hostel, my reflective mood continued with a long, intense chat with the proprietor. We talked about drinking and she related stories about her own experiences and those of friends, detailing just how detrimental it had been to their lives. There is no denying that part of the reason I felt compelled to make such a huge change in my life was the frequency and sheer scale of my drinking in London. Now, 10 months into my life-changing adventure, I noticed that more and more I was stopping to cool down with a beer and I didn't like the increasing frequency.

I'd been thinking about this a lot and recognised that alcohol was becoming a reward in my routine. Each day, on arriving at my destination, I would treat myself to a cold beer. My conversation with this lady confirmed to me that I would like to make some changes.

While in Quepos, I had heard about Flutterby House, an eco-hostel in the beach town of Uvita, so that became my next target. Uvita is the gateway to the Caño Island Biological Reserves and is famed for its epic diving, so was irresistible for me. Getting to Uvita was no picnic, however, as the humidity on the run from Dominical to Uvita reached new levels and even at 8am, the volume of sweat I produced was alarming. The forecast stated it was 30 degrees but would feel like 46. With the new level of humidity came new problems. Drinking about five to six litres of water per day meant needing to carry more and that meant more weight to push. I was also sweating so much that my clothes were constantly drenched, such that my shoes even audibly squelched as I ran. Runners might sympathise to read that the sweating was also causing a lot of chafing between my legs. Not only is that unpleasant and uncomfortable, but it also increases the risk of infection, especially when unable to shower every night, which was a threat I didn't want to relive after the torrid episode in Nicaragua. Efforts at blocking the sun were laughable, as the creams I applied to my skin would last mere seconds before streaming down my face and into my eyes. Electrolyte and mineral loss was a sizeable risk too, so to ensure I remained strong and healthy I had to make sure I was taking on nutrients and not just carbohydrates and sugar.

In spite of these challenges, I did make it to Uvita to find a chilled tourist vibe, with the main town being set away from the coast. Once there, my first task was to get myself on a dive boat for the next morning and luckily the one dive shop running the specific tour I wanted could squeeze me in. The second task was to find Flutterby house. It is entirely made from timber that has been found on the beach, which has been used to build tree-house dorms in the garden next to the main recreation area. From the eco-toilets to the daily communal meals, everything was about sustainability and community, and I instantaneously felt at home.

After an early night, I was ready for my marine adventure. Isla del Caño is a National Park about one-and-a-half hours by boat from Uvita so we left the beach at dawn.

Coral is not that common on the Pacific Coast of Costa Rica but Isla de Caño is an exception. Our first dive was to about 18m, at which depth we were immediately among the most staggeringly vibrant sea life. Immense shoals of wahoo circled above, white-tipped sharks slid along the sand, moray eels peered out of holes and in the distance a turtle glided past. More white-tipped sharks graced our second dive, along with an impressive sting ray and a spiralling shoal of colourful fish glinting in the sunrays above. Quite literally, everywhere I looked there were fish. Just to add to the already spectacular day, on our way back we spotted a huge pod of dolphins hunting fish. In the melee, they either skimmed along the surface or leapt straight into the air before splashing back into the ocean. Captivated, we followed them briefly before heading back to land.

The next morning, I rose early in a vain attempt to avoid the heat and humidity. This was a curious day because back in the UK the result of the General Election was to be announced so for the first time in a while, I listened to lots of news from home while running. This strangely didn't make me feel homesick but instead validated my decision to branch away in my own direction.

Despite the early start conditions swiftly became brutal, and it wasn't long before I collapsed at the side of the road seeking shelter from the oppressive heat. It was futile as my options were limited, so all I could do was keep running. Finally, on the outskirts of Palmar Norte, I stopped at a petrol station. Mid-procrastination, an older French gentleman handed me $20 to buy supplies. It never ceases to amaze me that these little gestures of generosity always happened just at the right moment, just when I was finding things physically or mentally demanding. This donation paid for a small room above the local restaurant and just in time because a rainstorm of epic proportions rolled through the village, turning the roads into rivers. As I sat there listening to the rain pounding the corrugated iron roof, it dawned on me that this was only the fifth time it had rained since I had left Vancouver.

It's easy for the days to blend together in my memory on my southward push towards Panama but some days stand out from the rest and the 60km from Palmar Norte to Rio Grande is one of them. Rainstorm over, I woke at four in the morning to get out on the road before the heat and humidity became too oppressive. This necessitated roaming the deserted streets looking for a water source to fill my bottles before venturing out into the countryside. On finding a tap round the back of the local bakery, I fuelled up on freshly deep-fried empanadas and orange juice and extracted as much cash as I could from the ATM. This day wasn't memorable because anything special happened, but because all the little special things, both good and testing, happening in succession. For starters, the scenery was amazing, like running through a tropical garden where flowering vines draped the branches of the tall trees that lined the road. Macaws, parrots, parakeets and vultures swooped overhead while frogs sang in the puddles. Bushes bloomed with red, orange and purple flowers that punctuated the verges. The hills were undulating but not too fierce.

This ravishing beauty couldn't prevent my stroller starting to buckle under the weight of the additional water I had to carry and at 31km a spoke snapped on a rear wheel. This forced a stop on the side of the road to carry out repairs. As I reached marathon distance, I heard a loud screeching sound behind me, followed by the sound of a car skidding. I glanced back to see a large four-by-four hurtling towards me with a startled driver fighting to swerve away from me. Luckily, they managed to pass me without disaster and I stood roadside with my heart beating itself out of my chest, and adrenalin surging through my veins.

After composing myself enough to start running, a second spoke snapped and forced yet another stop. Although this was frustrating and not typical, it gave me a chance to rest before embarking on the last 18km of the day.

At 50km, I was overjoyed to find a stall on the side of the road selling cold coconuts where the owner kindly refilled my water bottles. At this point, I had probably drunk between eight and ten litres of liquid. With the day coming to an end, I met a German couple, Robert and Sabrina, who were cycle-touring south and aiming for the same town as me. We arranged to meet up later in Rio Grande.

Just before arriving in Rio Grande, I found a river and let the cool water wash over my body. Eight hours and 60km of running in these conditions with the additional water pushing my stroller up to about 40kg had taken its toll on my body. As I laid there, I reflected on my time in Costa Rica, knowing that tomorrow would be my last day. It had taken me 12 running days to complete the 540km across Costa Rica and I managed to average more than a marathon each day, even with some magical breaks. Costa Rica is more developed than its northern neighbour, but both were immensely enjoyable in different ways. What stood out was the generosity and hospitality of the Costa Ricans, which had been overwhelming at times.

Just one more country lay between me and the halfway stage of my adventure. Crossing the border into Panama turned out to be far more complicated than others I'd encountered. It had a bureaucratic feel and involved going from one desk to another, which with a stroller was less than ideal, while the frequently bad-tempered staff made the whole process more frustrating.

Worse awaited me when I finally made it through into the kind of border town I had grown to hate. Shops crowded the small streets, sketchy-looking people lurked on street corners and there wasn't much in the way of affordable accommodation. At the sight of a Burger King sign, I unashamedly felt a craving to gorge on junk food. Unfortunately, nothing is that simple and before I could satisfy this urge, I needed to scour the streets for an ATM to obtain some local currency. Then, the Burger King was situated on the first floor, and I had to drag my stroller up the stairs just so I could sink my teeth into a Whopper and delay the inevitable search for accommodation.

Delaying tactics exhausted, I settled on a grubby hostel that was tacked onto a petrol station. With scant options for dinner, I made do with takeaway pots of ceviche and an unhealthy selection of junk food from the petrol station. After a trying day I retired early, feeling tired and irritable but determined to surface refreshed and in a positive mood to enjoy my first full day in Panama.

CHAPTER 17 - RUNNING FOR A PLANE

DISTANCE TO BUENOS AIRES: 8,455KM

Once in Panama, I only had 800km to run and over a month to do it. My first port of call was a town called David, about 50km away because I had been told there was a good hostel there and a restaurant that served the best fish soup. Small practicalities like these assume a disproportionate importance when you are not travelling for pleasure. Thankfully the Panamanian roads continued to be pretty good and I tried to cover as much mileage as I could while the sun wasn't too oppressive.

The Bamboo Hostel lived up to its reputation. With time no object, I took advantage of being at a proper hostel by booking in for three nights, helped by the friendly chap behind the counter. Beside the main attraction of a dorm in a treehouse, the hostel had a pool, a small bar, a kitchen and a very friendly coati that lurked around the fridge. This animal looked like a cross between a fox and a badger and spent most of the day walking along the fences, ever on the look-out for food.

Once settled in, I went in search of the amazing fish soup. As with many great dining experiences, the restaurant looked nothing like what I was expecting and was a small rustic establishment on the outskirts of a residential area. At mid-afternoon, it was closing up, but thankfully the chef was standing behind the counter and acknowledged me as I entered. He was possibly one of the largest men I had ever seen and wore an apron with food scraps clinging to the loose threads. Hygiene was obviously not a concern. I asked if food was available and he looked unwilling until I mentioned the fish soup. His demeanour rapidly changed and he waddled over to a large pot to start ladling some into a bowl.

The fish soup was to die for and most definitely the best I have ever eaten. It was thick in texture and strong on taste and, accompanied by the chef's homemade chilli sauce, it was incomparable to anything I had tasted before, though the 50km of running may have had something to do with that.

Belly nicely full, I returned to the hostel and set about my routine. Afternoon became evening and I mingled happily with the other guests over beers. The next day was devoted to lounging around the hostel and catching up on expedition admin, while a plan was devised to throw a barbecue that night. Staff and guests grouped together to put on an absolute feast, and we bonded over drinks and travel stories.

By my third day in David, the urge to get back on the road was unmissable. Spending any more than a day in any particular place was not normal, and while I enjoyed the contrast between my life and that of the backpacker, I could only endure it for short spells.

The next morning, although eager to get back on the road I was less enthused by the roads I was running on. At this time the Panamanian Government was enlarging its section of the Pan-American Highway, which meant heavy roadworks, with traffic squeezed into tighter lanes and very little room for a small stroller-pushing runner. The going was hard and it was frustrating to continually have to jump to the side of the road to avoid the traffic. Worse, the rocky ground made running with the stroller nearly impossible and the going was painfully slow. Feeling tired and with flagging motivation, I pushed on and eventually reached the town of San Lorenzo.

This part of Panama is quite agricultural with lots of flat, verdant land on either side of the road and huge trucks speeding past, which was unnerving at times. On the outskirts of San Lorenzo was a large restaurant and shop next to a petrol station. On enquiring about accommodation and being told there was nothing nearby, I was granted permission to camp behind the service station, but only at dusk.

That night I slept so well. I hadn't been in my tent for a while and was surprised by how refreshed I felt. The only downside was that the food at the restaurant hadn't agreed with me and once again I was in an urgent rush to find a toilet. The one in the restaurant was locked, prompting a desperate search for someone with a key. Finally, one of the guards appeared. He reluctantly handed over the keys and I possibly unsuccessfully tried to hide my relief as I dashed to the toilet with more than a touch of urgency.

I had to decide how I was going to proceed. Having a bad stomach brought several complications, including being very low on energy and missing valuable minerals. It might also prevent me replenishing them later in the day. To add to all this, it was incredibly humid which would only compound the difficulty of staying hydrated. My choices were either to remain at the petrol station for a day of rest or push on to San Felix, 35km further south. With the number of rest days I had taken recently weighing on me and despite having a valid excuse, I decided to push on. Fortified only by two jam sandwiches, the most inoffensive meal I could rustle up, I hit the road.

Starting with a walk, I did not get more than 2km before needing to search for an appropriate place to jump into the undergrowth and do my business. At these times I just had to leave the stroller unguarded on the side of the road and hope that someone didn't take advantage of me literally with my pants down. To make matters worse, the humidity and lush surroundings made hedges the perfect environment for biting insects, resulting in the whole affair being even more tortuous.

After another 10 more kilometres, that dreaded tightening feeling gripped my abdomen. Pins and needles tingled all over my body and I could not ignore the signs. The next problem was the lack of restaurants or petrol stations while being in a built-up area.

Clenching every muscle in my body I made my way gingerly onwards, determined to find a suitable spot. After about 2km a small roadside restaurant came into view. I deserted my stroller, removing my valuables and scurried to the bathroom. The floor was covered in filth and in the cubicle a bin was overflowing with used toilet paper and even an empty bottle of rum. Cruelly, there wasn't even a lock on the door. On the plus side, there was a toilet seat, a luxury that is not always provided. The potential disaster was averted, for now.

I emerged a new man and hit the road. Weak, but happy, I got to the village of San Felix, after detouring for a few kilometres off the Pan-American and up a steep hill. Here, the air was close and rain loomed, so finding digs was a priority. San Felix is home to the Hospital Regional de Oriente and as a result, the streets were lined with ill-looking people being escorted by their families. There was an odd symmetry to landing here after the preceding day, and I hoped this place of healing would nurture me. Finding a small hostel, my main priority became eating enough to power me through the next day. Alas, more drama was in store. As I took the first bite of my sandwich I felt the crown on one of my teeth dislodge. I sifted through my food and managed to extract it. Luckily, apart from the visual difference, it didn't cause too much of an issue as the nerve had been extracted so I felt no pain. I took this as another indication that my body, as well as my mind and stroller, were feeling the effects of the 9,000km I had already run.

The next morning, the landscape began to morph into hills covered with dense green forests, while a road below swept through the small valleys in between. While the running was harder, the feeling of being back in better sorts made me more determined to push on. On the sections of road where roadworks weren't causing havoc, small huts opportunistically sold brightly coloured hammocks, clothing and bracelets.

I had a feeling that finding somewhere to camp was going to be difficult due to the dense woods and roadworks. With luck, as the clouds threatened to burst, I came across a small church on the side of the road. It was about 5pm and the evening service was about to start. Seeking out the person in charge, an elderly lady, I asked if it was possible to set up camp. She amiably indicated a small adobe hut to one side of the church. At any other time this small, partly open-sided building with a dirt floor would have seemed like a hovel, but today it was a palace. There was a bench where I would be able to cook and a blue plastic chair to sit on. Paradise! My new host returned to the church while I set up camp to the strains of Spanish hymns. After the service drew to an end, the second priest of my journey came to introduce himself and listen to my story, and even wanted a selfie with me! When I started this expedition, I had packed a couple of emergency Firepot dehydrated meals. That night I ate the first of these before curling up on the mud floor.

The town of Santiago lay 70km to the east and having endured both running with a bad stomach and putting in 131km over the last three days, I decided that if I could make the distance I would deserve a day off there. A little research produced evidence of a small hostel in the northern part of the city, so I aimed there. I did arrive in one piece, although was taken aback by so much American influence, something I hadn't seen for a while. The main road into the heart of the city was lined with KFC, McDonalds and Pizza Hut outlets. The traffic was also much denser than I had been used to. My guard instantly went up in this town as I navigated its small streets in search of my hostel. In a surreal scene, a school band rehearsed on the main street, oblivious to the passing traffic. As with a lot of Central America, the band was made up of drums and trumpets, fronted by a baton-wielding troop leader doing somersaults and aerobatics. Cars would slow down to admire the practice before moving on.

Finally locating the hostel, I was prematurely relieved as my body was weakened by the 70km I'd just run. So far, this was the furthest I had run in a single day and it had taken between nine and ten hours due to the terrain and the weather, plus the mental and physical toll. Exhausted, I rang the doorbell and waited. In the eerie silence, I walked around the property, peering through the metal fencing looking for any sign of life and shouting, 'Hola', each time louder and more urgently. No one was there and the rain was starting. Recalling this moment is just as vivid now as this was a huge blow. My body was spent and here I was standing in the rain with nowhere to sleep and only the vestiges of energy to do anything about it. Painfully slowly, I found a hotel near the bus terminal that hugely was out of my budget, despite the discount I was offered. With no other choice, I just needed to collapse and get clean and when that was done, I targeted McDonalds to consume as many calories as possible.

Happy to leave Santiago, Aguadulce, which translates as sweet water, sounded like a good place to head for and after resting for a day, it was within reach at 60km. The road started nicely with awesome views down through little valleys but by lunchtime, it had reverted to a dual carriageway, with larges lorries and roadworks to dodge. I could feel the interest and motivation draining away and the number of rest stops started to increase. Not one to give up, I was spurred on by the people who stopped to talk to me and provide the impetus to struggle on.

Despite the intense humidity building along my route, I made the 45km to Penonomé quite comfortably. The rain had returned for the final few kilometres, making it unpleasant while I searched for somewhere to stay. In a seemingly unremarkable town, the hotels here were all too expensive for me, leaving only a small place a bit further south. Warnings about this place rang in my ears and when I arrived I could see why. The building looked halfway between derelict and under construction, with barred windows and a 'Se Vende' (for sale) sign hanging from the unfinished windows above. Everything about it seemed wrong but with more rain imminent and over 45km in my legs I needed to shelter and get cleaned up.

I found the lady in charge and managed to secure a room for the night. Although cheaper than the places in town, the price still seemed expensive for its appearance. The interior reminded me of horror films where tourists are taken prisoner and tortured. Wires tumbled from the ceilings of unfinished corridors, the doors were all different sizes and there was an unsettling echo. However, the windowless room itself had a clean bed, plasma TV, and a bathroom with a shower complete with a shower head and a decent toilet, two things that are not a given in this part of the world. The lingering smell of disinfectant was a good sign, although I was stuck with it as there was no way to air the room even if I wanted.

Uneasy but settled, I ventured out for supplies, with hair products high on the list. Over the last 10 months my hair and beard had grown into a shaggy mess. Passing the cosmetics counter I noticed some individual sachets from a company called Nevada Natural Products, on which shampoo was written in large letters. Delighted, I got back to my hotel room and jumped into the shower, opened the packet and emptied it over my head. As I scrubbed my scalp and beard it foamed up beautifully and I enjoyed the sensation of getting clean again. Yet once out of the shower again I looked at the packaging more closely and noticed the ingredient "esperma de ballena". It dawned on me that this could be translated as whale sperm. Had I just washed my hair in whale sperm? Some investigation revealed that it was spermaceti, a wax that is most often found in the head cavity of sperm whales and dolphins. There was a moment of disbelief and horror in which I started to wonder just how legal or ethical this was. Further research only made me realise just how hard it is to understand what is happening in the whaling world, and how widespread these ingredients still are, which is alarming. Spermaceti is not legal in the US or the EU.

After this fairly joyless episode, the seaside of El Farallon del Chiru was exactly what I needed and being so ahead of schedule meant that I could afford to spend a few days for some important downtime. It wasn't the physical side of the journey that was wearing on me but the relentless travelling day in, day out. My routine was so entrenched that I was just going through the motions each day to get the kilometres done. The problem was neglecting the edifying aspects of the experience, and processing just how big an adventure I was undertaking. It had become humdrum. The antidote was a perfect little hostel, run by a French couple, where the guests were people I could properly relax around.

The next three days were spent relaxing, walking on the beach and eating as much food as I could. Admittedly, the beach was not the most idyllic, with dogs and vultures fighting over fish carcasses discarded by the local fisherman. But it was tranquil and so all that I needed at that time.

It was 22nd May and my flight to my sister's wedding was booked for 19th June. My original plan was to run to Panama City and board my flight but with less than 150km to run and nearly a month to cover the distance, I started to investigate other options.

Feeling more grounded and physically restored, I hit the road again with Panama City only three running days away. Europcar billboards became my guides to where to stay, as they gave the distances to the next towns that coincidentally matched my range. My rationale being if there was a car rental shop then there were probably the essential amenities I required. Chame became my next focus.

I got there on a Sunday and everyone seemed to be drinking. Three men were engaged in a fight with some girls in bikinis, but with the girls very much in control I decided to leave them to teach the men a lesson. Later that morning I met a toothless chap on the side of the road who was selling mangoes, albeit in a rather unproductive manner. He got very excited when he heard I was from Scotland and commenced filling my stroller with as many mangoes as he could. His excitement was down to the fact that he knew someone from Scotland. On this basis, he decided that we should call his friend that very moment and tried to persuade me to leave my running stroller on the side of the road so we could call from his house. To extract myself from this increasingly weird situation, I politely thanked him for his kindness, paid for the mangoes, and backed away.

By lunchtime I had arrived at the small town of Coloncito, a more touristy hang-out where surf shops, cafes and small souvenir stalls crowded the streets. The ice cream parlour proved too much of a temptation for me and a useful place to research the nearby hostels. One on Malibu Beach popped up and I decided that the detour was worth heading off the main road along a road that petered out into a mere trail. The road meandered through new gated housing complexes and the small beach town of Neuva Gorgona. Nothing seemed finished on this part of the beach and I started to fear my detour may have been in vain. Large, grand houses lined the beaches and nowhere did it look fitting for a surf hostel. Eventually I matched the property with the advertised hostel and it looked decidedly shut. Waiting around outside looking for signs of life eventually prompted a greeting from a lovely lady who spoke fluent English. All the while a family was packing up and I couldn't work out if they were guests or relatives. After chatting, it transpired that the property had been a hostel but the people managing it had been doing so inappropriately, hosting large parties and other indiscretions. This lady was the owner and she was here to turn it around herself. Her first gesture was to arrange a huge jug of iced tea for me and to inform me that I was more than welcome to stay.

My host was incredibly kind and made sure I had everything I needed, including food for the evening and the next morning. Perhaps she was happy and a little comforted in a tricky situation to have some company and we talked about Panama, the complicated politics and what was going on in the capital. The longer I spent in this country, the more I sensed that everything was a lot more complicated under the surface. Recent times had been turbulent and apparently there had been a lot of fraud in the Panamanian government. Panama has always had a bit of a strange past due to the American-controlled canal, as well as contending with friction between native peoples versus colonial powers. To top all that, there is a very wide gap between the country's rich and its poor, which engenders considerable dissent. These issues perhaps explained some of my stranger encounters.

The next day there was nothing I wanted more than to get within striking distance of Panama City. It didn't seem to matter that I had an abundance of time, Panama City had occupied my thoughts for so long that I was itching to get there. With only 85km between us, I was only two days away.

I rested in La Chorrera. Alas, this was not the most inviting town, with stray dogs sleeping on every corner and dark doorways occupied by sketchy looking men. While the place was depressing, I was in surprisingly good spirits. Tomorrow I would arrive in Panama City, a city that had been the focus of every day for the last 283 days. I had run over 9,000km, pushing everything I needed to survive and tomorrow would mark the completion of that goal.

When I first started, I didn't let myself contemplate what I had to do to get there but now on the eve of arriving, I once again dared to dream that this whole expedition was achievable. Of course, there were many challenges and obstacles ahead but everything I had overcome thus far imbued me with a sense of invincibility.

The next day was much like any other until I surmounted the brow of a hill and got my first glimpse of Panama City just 15km away. Happiness overwhelmed me, generating an enormous energy boost. I met a fellow traveller, a Russian on a bike, who must have been mystified as my sense of achievement bubbled over while closing in on the final obstacles between me and a few days off.

The first of these obstacles was the Bridge of Americas that spans the mouth of the Panama Canal. This iron bridge is made up of four carriageways and no hard shoulder but with so little distance to cover I threw caution to the wind and took my place among the lorries and cars as they motored south.

After successfully navigating the bridge I escaped the busy highway, passed the Estadio Maracaná and found a footpath along the seafront. Getting to the hostel I had booked required me to pass through the ghetto part of Panama City's old town. Various people had warned me of the potential risks in this area and advised me to steer clear, but with so few kilometres to go, I couldn't resist one final thrill. The formerly magnificent buildings had been left to crumble and fall into disrepair and the once-bright paint had been dulled by damp. Doors were blocked with bricks, and windows hid behind iron bars between smears of graffiti and the satellite dishes that clung precariously to the brickwork. While everyone I saw stared back at me, I never felt in danger. More a sense of sadness that in a city of such wealth and foreign influence, places like this still existed.

The team at the Luna's Castle hostel checked me in and showed me where I could keep my stroller while staying there. I found my dorm room and cleaned myself up sufficiently to visit local bars, drink beer and devour sushi in chic and clean modern restaurants. It was the complete anathema to everything I had experienced on the road.

That evening I joined a group of travellers who were sitting around dissecting their travels. Some of them had been to amazing places but listening to the words, 'I did the ruins' or, 'I've done Bolivia' made me squirm. As a backpacker in 2007, I'd been exactly the same. My itinerary was dictated by ticking off as many places listed in guidebooks as possible. You had to visit here, stay there and eat this. Listening to everyone's conversations made me smile inwardly at just how much I loved my current adventure and the freedom it imparted. I may not have visited all the ruins or seen all the sights, but I had experienced a real cross-section of a whole continent. The process of entering at point A and leaving at point B made the straight line between the two nothing more than connecting dots. I went through the affluent areas, the poor areas, the dangerous areas and the beautiful ones. Truly, I had experienced more of each country by not having a guidebook and instead travelling through the different communities, meeting real people and experiencing real life.

I had not even been 12 hours in Panama City, but I was already thinking about what to do with all the time before my flight. While I had originally planned to stop in Panama City and fly to Cartegena in Colombia, all the maps showed that the road heading southeast from Panama City to Yaviza, the gateway to the infamous Darién Gap, was just over 300km. This was eminently achievable with the time I had spare and the idea of delving a little further into the Darién region filled me with excitement. So, decision made, I resolved to extend my run south.

Even with this extra running I would still have lots of time in Panama City. While I really didn't want to kill time in Panama, my budget didn't really stretch to the $400 it cost to change my flight, especially as I still didn't have enough funds to get to Buenos Aires.

Needing company, I joined the other travellers to share stories. Conversation soon gravitated to my journey and people sat around asking questions, trying to piece together what it was that I was doing. I explained in detail about my trip and even my current conundrum. Later that morning a girl who had been sitting at the table with us earlier approached me. She was a pretty Canadian called Jen with eyes that sparkled and a huge smile. She pulled out an envelope and handed it to me. 'Open it when you are in your room,' she said then disappeared. I was confused but thanked her and went straight to my dorm to find out what was in this mysterious envelope.

On prising it open I was astounded to find $400 and a note that simply told me to spend more time with my family. This amazing act of generosity blew me away, especially from someone who had no idea who I was. My immediate reaction was that I couldn't accept it. I was not broke, I just didn't have the extra money I needed since my schedule had shifted. Straightaway, I sought Jen out and explained I couldn't take the money but she was adamant. Stunned, there seemed no other option than to do what she told me. At my computer I booked my flight for 10th June, a whole nine days early. There is no way to express how grateful I am to Jen for what she did. It added such a beautiful end to this first half of my adventure and I will never forget that act or her.

Now, less oppressed by time stretching out before my flight, I explored Panama City as well as the ghetto area but this time armed with a camera. I wanted to capture the beauty of the crumbling buildings, despite their current condition. I walked up and down the dilapidated streets and absorbed as much as I could, including a very strange encounter with a local couple who were sitting in a doorway and evidently under the influence of some narcotic. 'Would you like to take some photos of my girlfriend naked?' he asked, as she nervously looked up at me.

I stared at him in disbelief. 'You can go inside and take photos of her,' he continued. There was no mention of money, though I am sure that was assumed. Slightly taken aback I politely declined the offer, wished them a good day and made my exit.

In the fish market between the new and old towns of Panama City, I found a small table outside one of the eateries and ordered some fresh seafood and beer while I sat and watched the diverse crowd that frequented the market. There were the local fishermen, those who worked in the bars and on the stalls and then the customers, ranging from the affluent to the poor with a handful of other travellers mixed in. My moments as the fly on the wall watching the world go about its business were somehow so comforting. Even though I was still alone, this was solitude in numbers.

My last tourist excursion was the obligatory visit to the Panama Canal. While in Panama I felt I should probably visit one of the biggest engineering feats of the modern world. The canal itself, while impressive, was less than inspiring to me. It's not clear what I was expecting but it was just a really big canal. However, the museum that accompanied it was informative about how the canal was built, plans for its future but more interestingly, how it impacted and shaped the country. The canal itself is 82km long and connects the Atlantic and Pacific Oceans, meaning large container ships don't have to make the long and perilous journey around the Cape Horn. The Americans took over construction after a failed attempt by the French and it finally opened in 1914. The total cost was about $375,000,000 but it also claimed nearly 6,000 lives, most of which were those of indigenous people. The Panama hat, made in Ecuador, was designed to protect the workers on this project from the oppressive sun. Interestingly, in 2013 the Nicaraguan government awarded the Chinese a controversial 50-year concession to build another canal, the fear being that the US exerts too much influence over the only existing route.

Having spent three full days eating, drinking, and being a tourist, it was time to hit the road again. With nearly 300km to cover and a route that ended in the mysterious Darién Gap and was relatively untravelled by tourists, I was itching to get moving once again.

CHAPTER 18 - AN EXTRA 300KM JUST FOR FUN

DISTANCE TO BUENOS AIRES: 7,968KM

After Panama City, I yearned to get back to the more rural parts of Panama. On the grapevine I'd learnt that an earlier relic of Panama City lay to the south of the capital. Panama Viejo, or Old Panama, was the original site of the city until 1671, when Welsh pirate Henry Morgan ransacked and completely destroyed it.

It was on my way there that disaster struck. While running the front wheel broke away from my stroller and rolled away. I chased it down and discovered that the axle had clean snapped. Stranded in a fairly questionable part of town with no way of either advancing or retreating, I approached a young mechanic and tried to explain what had happened and why I was keen to get it fixed as soon as possible. He took one look and immediately knew what had to be done. After rummaging in a box of scrap metal, he produced a piece that matched and proceeded to weld them together, before verifying that it was sound. Within 40 minutes my stroller was back on the road and I was pushing it towards the Darién Gap. My admiration for the mechanics here knows no bounds because they never look at a problem and think the worst; instead they just see a solution and somehow work magic.

I don't know how far I intended to run that day but it certainly wasn't 62km. My route wove past the airport and out of the city, through the ugliness of kilometres of industrial buildings.

I then had my first interaction with a Panamanian police check. The officers in charge wanted to see my passport but more importantly, also my assurance that I was not aiming to cross the Darién Gap. They needn't have worried, as I explained that I had a ticket back to France in a matter of days, which they accepted and ushered me through.

My next issue was finding somewhere to sleep. With nowhere to pitch a tent by the side of the road, I pushed on to Chepo. A traveller I had met in Guatemala had told me that he'd been very lucky with fire stations. Apparently, if asked they are more than willing to take you in so I looked on the map and found a fire station. But after a few days off, my legs were starting to struggle and it took all my strength to get there. When I did, the fireman I asked about camping looked a little perplexed and went off to find a superior officer. The next fireman I spoke to seemed more understanding but informed me that there was nowhere I could stay on site. However, this was the new station, and there was an old station across town that was no longer in use. The key would be getting there before its guardian left. As tired as I was, the idea of a free roof over my head was too enticing to pass on. Without hesitation, I ran through the centre of town in search of the old fire station.

What I found was a dilapidated old building with junk piled on the forecourt but also a smiling guardian. He showed me into an interior that was in a similar state to the outside. The kitchen was basic and dirty and the living quarters were a muddle of bunk beds and old, damp mattresses. But it was essentially dry and secure and that was all I needed. There was even a shower, of sorts, and a toilet and while it wasn't anything to brag about, it served as home for the night. I set up my camping area in a space on the floor, cooked some dinner and rested.

The next day I felt pretty wiped out and trying to get into a rhythm was hard. At a small village about 20km along the road, I was close to calling it a day, but my new hard deadline of 9 June propelled me onward. The humidity was stifling and the further I went the more rural and exposed it got. By mid-afternoon I had covered a mere 40km and arrived at the bridge over Lago Bayano, where a cluster of huts around the road sold knickknacks and provisions to passing cars. I made it to the other side where a man was sitting next to bunches of plantains. To his right was a large wooden settlement that looked ideal for camping next to. 'Excuse me, do you live in this building above the lake? If so, can I camp next to it, as it is safer than being alone on the side of the road.'

Unflustered, he looked me up and down before replying, 'I do live there and you can camp but you must wait until I have sold all my plantains.' And with that, he sat back down.

The art of selling plantains seemed to entail nothing more than sitting next to plantains at the roadside and hoping cars would pull over. Time crept by slowly and as the afternoon rolled on the pile slowly diminished until it was time to go home. With no small relief, I followed my host across the road and up a dirt track to the building.

It turned out that I had stumbled upon a group of the Emberá Tribe who had lived in this part of Panama for hundreds of years. The man also indicated that Lago Bayano is not a natural lake, having been flooded in 1976 as a reservoir and named after the African slave, Bayano, who led the biggest slave revolt of sixteenth century Panama.

These kind people offered me a small patch of grass away from the main building on which to set up camp. One of the kids gleefully helped me and was fascinated by all my kit, especially my tent. Once set up, my host came and asked me to join his family for a bath in the lake. He led the way, followed by his heavily pregnant wife, his three children and me down a perilous mud track to the water. Astonishingly, his wife made it down the steep, slippery bank easily. Once in the water, we all set about washing and a communal bar of soap was passed around. The kids enjoyed having this strange white visitor and we had a lot of fun splashing around. They asked if I was hungry and within seconds one of the boys was up a tree and knocking mangoes down to his brother. It couldn't have been more perfect. Just before we dried off, I went under the water and came up slowly with my long hair and beard straightened. One of the young boys looked over at me and exclaimed 'Jesus!'

Later that evening I was introduced to the others in the settlement, including the chief and invited to join them for dinner. Questions were fired at me about London, the way we live and the differences between our lives and theirs. It was good to offer something in return for their hospitality by ending the night teaching the young children English using the textbooks they had in the house.

The next morning I said my farewells to my hosts and hit the road south. In contrast to the day before, I was filled with huge optimism and happiness. The remote road here was exactly what I had dreamt of before leaving Vancouver. It was humid and surrounded by dense jungle. Even after nearly a year on the road, I was still craving something more adventurous, more challenging or even just new. Part of me had wondered if the thirst for adventure might dry up, but that never seemed the case. The more I journeyed the more adventure I sought.

This particular morning was like no other. In the stillness of the jungle, the calls of the howler monkeys rang from the trees and halted me in my tracks. Volleys of calls and responses were exchanged across the dense forest. Unable to resist the chance to engage them in conversation, I attempted to join in. Alas my calls did nothing more than trigger warning messages from the monkeys that a human was encroaching on their territory. In my mind, I thought we were getting on famously!

The roads straightened as they cut through the rainforest, which should have helped, but in fact the going was slow because lorries had churned up the tarmac. This was because they ferried logs in large quantities away from the rainforest, which was being destroyed on a vast scale.

Some of the locals I asked told me that the indigenous people were now expected to pay taxes so had to find a way to raise cash. According to them, the quickest fix was to chop down the trees and sell the wood, then use the newly exposed land for grazing beef cattle. The devastation was huge, with large swathes of rainforest turned into brutal scars on the land. The stilted houses of the local tribal villages were now exposed to the road as signs of the modern world crept in. The old world was being swallowed by the new and the forest sacrificed to cater to our needs and wants. It was depressing to see the effects of capitalism on even the remotest of places.

There is always something exciting about entering an area that is perceived to be treacherous. My route contained many places that people warned me about or expressed concern about me entering. In America it was Mexico, in Mexico it was Guatemala, in fact, nearly every country maintains a negative impression of the country to its immediate south and I often wondered if the reverse would have been true. Would everyone have raved about their exceptionally friendly neighbour to their north? In each of the regions I ran through, certain incidents or groups of people have caused people to speculate and spread rumours of danger and tyranny. The Darién Gap was certainly no exception.

The Darién is the thin sliver of land that connects Central and South America, or what is today the border between Panama and Colombia. The Gap refers to the only missing piece of the Pan-American Highway that, except for this 100km, runs unbroken from the north of North America to the southernmost tip of South America. There is more water here than land, with numerous meandering waterways penetrating thick tropical jungle. It's easy to see why nobody has even attempted to build a road through it, though many plans and schemes have been proposed.

Perhaps what most appealed to me was that the Darién has a Scottish connection dating back to 1697. It was chosen as the recipient of one of the earliest public subscription schemes, the Company of Scotland, better known as the Darién Scheme. Founded in 1693, the Company raised the equivalent of $47 million in today's money, which amounted to over 25% of the wealth in Scotland at that time. Thousands of ordinary people subscribed to the Company, which planned to create an overland route for trans-shipping goods from the Atlantic to the Pacific, thus avoiding the treacherous and long trip around Cape Horn, a very forward-thinking notion for its time. The expedition left Scotland in 1697, but the planned self-sufficiency from the crops they planted failed to materialise.

Disease, hunger and Spanish settlers turning against their Caledonian neighbour all combined to help end the scheme with almost 2,000 of the 2,500 settlers losing their lives. Scotland could not survive the loss of such an enormous proportion of its national wealth so persuaded England to pay off the national debt in exchange for the Acts of Union that united the two countries in 1707. The indigenous peoples, Kuna and Emberá, constitute the majority of today's population but a Scottish influence can still be seen in some place names.

The fact you have arrived in the Darién Region is impossible to ignore, thanks to the large signposts that frame the road and the increasing presence of the SENAFRONT. In 1989, the US invaded Panama, prompting the country to dismantle its military forces. The SENAFRONT was formed as the national border force and has a specific remit to patrol the border with Colombia, which experiences a lot of illegal immigration.

The first proper checkpoint is between Torti and Santa Fe. All traffic through the Darién must use this checkpoint where every person is screened and questioned. As with most of the military I'd encountered thus far, the SENAFRONT did not know what to make of a runner pushing a baby stroller. The first officer who approached me rather gruffly motioned me to the side of the road. He instructed me to leave the stroller there and report to the officer in charge. Most of the work is conducted in a hut on the side of the road but special cases are invited or, more accurately, escorted to the main office, which in this case was in a run-down building set back from the road. That day, I had clearly won the special lottery, so did as I was told. A lady wedged into a chair gave me a disapproving gaze that said, 'you've disrupted me; you better have a good reason.' I explained my mission and she looked at me in utter disbelief. At the beginning of my journey, it had been harder to get officials to take me seriously but with the stamps of eight other countries in my passport, my story had become more credible. Her first question was how far I was intending to go. From her tone, I could tell she was used to bold and sometimes foolhardy tourists aspiring to tackle the Darién Gap. I assured her that my intentions were limited to getting to the end of the road at Yaviza and that I was very much set on getting a bus back to Panama City.

We went through the motions of trying to make the whole process as official as possible. Thumbing through my passport, she asked more questions before directing me back to the main post at the side of the road. My passport was studied again and then various details were recorded on a scrap of paper. The last part of the security check was the customary stroller inspection. Whenever an officer pulls back the plastic sheet and sees each waterproof bag, individually packed and slotted into position, they seem to lose interest almost immediately. The wave through follows and I am unleashed once more onto the open road.

From the checkpoint to Yaviza was three days' running. My first night was going to be spent in the community of Santa Fe before aiming to camp at an eco-hotel for the second. I had done some research on camping options and a hotel called Canopy Lodge seemed perfectly positioned. At over $100 a night my budget wouldn't stretch to a bed but the owner, Raul, had replied to my request to camp and by some quirk of fortune said that he would be there at the same time.

A storm was brewing and it was chasing me down the road. If it caught up, the road would become impassable with my stroller, so the clouds building behind me were a real threat. I knew where I needed to get to but with the road stretching out in front as far as the eye could see, was unsure I had the mental strength to keep pushing. The day had a nightmarish quality, in that although I ran and ran, I seemed to be getting nowhere. My disappointment only intensified as the town on the map never seemed to appear. In another blow to my morale, a sign signalled that Santa Fe was in fact a few kilometres off the main road, but with rain on the way I simply had no choice but to dig deep. More than 50km into the day, I just wanted it to end.

The next morning, on my way back to the Pan-American Highway, a small boy accompanied me on his way home. As we chatted in English, his reaction to what I was doing lifted my spirits. The idea of travelling was so alien to him, nor could he compute the fact that I was doing this out of free will. I pressed on south towards Meteti and made good ground as some of the humidity had evaporated after the previous night's storm, which made the going somewhat easier. Meteti was an interesting little outpost and the last proper conurbation before Yaviza and the end of the road. There was no need to venture into the town itself, so instead I rested at the large warehouse-style supermarket by the road. Its covered foyer seemed a popular meeting place for the locals, many of whom appeared to live deep in the jungle and had come to make their weekly shops. With all the deforestation and new farming, new money was evident, widening the gap between rich and poor as Cowboy-esque farmers mingled with poorer indigenous people.

Running south from Meteti, I was joined for a brief section by some young, shoeless kids who tried to keep up as they gripped shopping bags. Raul had arranged for a chap to meet me at the side of the road in a new pick-up, so I dismantled my stroller as much as possible and together we loaded it onto the back. Canopy Lodge targets keen birdwatchers from all around the world, so the pick-up had been designed with passengers with binoculars rather than baby strollers in mind, but after a few attempts we managed to fit it in and were soon following a rugged path into the jungle.

Raul greeted me on arrival. He was a thin, friendly-looking older man who was dressed in simple khaki-coloured clothes that were perfect for this environment and exactly what you would want him to be wearing. On enquiring where Raul thought I should camp, he just smiled back at me. 'This is a quiet time of year and I only have one guest from China. You will be staying in one of the lodges.'

After my arduous journey, I was completely bowled over by this amazing good fortune, but Raul kept speaking. 'You really should stay for a couple of days and have a rest. The deal is you eat all your meals with me, it's no problem.' I didn't know how I had deserved such generosity but was not going to refuse such a generous offer.

The amazement continued when I saw my bedroom, which was about as far from my normal sleeping arrangements as one could imagine. Each luxury tent was situated on a beautiful wooden platform facing the jungle. Inside was a sumptuous double bed with bright white linen and everything you would expect from a five-star hotel. To add to the experience, an outdoor ensuite bathroom afforded amazing jungle views. Everything had been designed to deliver rugged luxury while quietly blending into nature, and this theme continued throughout the hotel. The main building was completely open to the elements and housed a dining room, lounge area and a library of books on both the history of Panama and the local birdlife.

Over the next two days I let this oasis nurture me, enjoying delicious food and taking walks through the dense jungle. The hotel's bird expert was on hand and took pleasure in seeking out rare birds such as the Black Oropendola and calling them over to where he was standing. From where I stood I saw only trees but his trained eye saw a whole different world and it was a rare treat to share it with him.

Another highlight of the stay was a visit to the local ranger station. This building had been deserted for some time and Raul had decided to restore its original function—to monitor illegal poaching and deforestation. This project, wholly financed by Raul, would ensure that the station would be occupied 365 days a year and was a great example of a private enterprise investing back into its local community to ensure the conservation of the land and the natural wonders it nurtures.

Tearing myself away from Canopy and after reiterating my gratitude to the team there, the final stretch to Yaviza was filled with mixed emotions. This would be my last day of running for nearly a month, due to my sister's wedding, and part of me wanted to savour it for as long as possible. At the same time, this was such a huge landmark in my expedition that I was eager to finish strong and be proud of what I had achieved.

Yaviza was an unknown quantity, and the fact that it was the last town on this section of the Pan-American Highway lent it an extra air of mystery. I knew the road was going to end abruptly, and that I would have no choice but to stop. On cresting a hill, I could see the outskirts of the town below, where makeshift homes crammed together around muddy paths. Slowly the town started to take shape and as I rounded the last corner, Yaviza came into view.

In my mind, there would have been some definitive end to the road, a sign or a plaque like at Land's End in Cornwall, but there was nothing. The road unceremoniously petered out, slowly morphing into a side street that disappeared into the backstreets. Alone on a dirty street, a wave of disappointment rushed over me, instead of the euphoria or sense of achievement I might have imagined. It hit me that the joy of this expedition actually lay in the day-to-day adventure rather than any success of finishing. Desperate for a reaction from someone, I called my mother and tried to sound excited about being at the end of the road. She congratulated me, then quickly switched subject to some dinner party she was going to. Standing alone in such a remote part of the world, having completed something so extraordinary did not feel like I thought it would.

Deflated, I switched into post-run mode and asked several people about possible rooms for the night and as expected, the options were very limited. There could be no anonymity here, as I was also advised to check-in with the SENAFRONT just in case anything happened. To get there meant negotiating uneven dirt streets where run-down houses were propped up on wooden stilts.

The SENAFRONT base was protected by an officer whose attention to protocol rather conflicted with my desperation to lie down and recover from the day's run. He insisted that I leave my stroller outside on the road rather than wheeling it inside the gate to a more secure location. My initial protestations were clearly falling on deaf ears, so I relented, gathered my most precious possessions and went inside. The barracks superficially resembled an army outpost, with sleeping quarters, flagpole, and khaki-covered kit, but instead of a parade, a group of 40 civilians from all different reaches of the world sat in a circle under the watchful eye of numerous SENAFRONT soldiers who pointed machine guns in their direction. I was taken to the very process-driven administration officer at a desk. I answered all his questions, eager to wrap things up as soon as possible. While he was inspecting my passport and filling in forms, I noticed a large whiteboard behind him. The board was split into a grid on which was a tally.

If you look up the Darién Gap on Wikipedia, there is a section on crossing the Darién. It talks of the Trans-Americas expedition of 1972 in which James Blashford-Snell and his team crossed in Range Rovers. Three years later in 1975, Robert Webb crossed on a BMW motorbike and there have been other notable crossings on foot. Crossing the Darién has always been a symbol of the most daring of adventures. So, it was something of a surprise to see this tally of the numerous illegal immigrants crossing the Darién Gap in a quest to reach the United States. The 40 men sitting on the ground outside had been recently rounded up. Were they any less daring than explorers with support teams and top of the range vehicles? From what I could see, the weekly tally was over 300, including representatives from Nepal, Ivory Coast, Somalia and about 10 other countries. Each had fled their country with nothing, many made it across an ocean and then walked across the Darién with no special equipment, no support team and probably very limited provisions. To cross successfully you must deal with snakes, disease, drug traffickers and the 57th Front of the Revolutionary Armed Forces, a rebel unit based in the Darién. The terrain is uneven, the route unclear, and this is regarded as one of the hottest and wettest places on earth. To put things into perspective, it took the first all-land auto crossing 741 days to cross 125 miles.

All this made me reflect on my expedition. I gave up a good career in London to undertake an exciting expedition over two continents, putting myself in danger purely for the thrill and yet here was a group of people who had risked absolutely everything to find refuge and potentially a job. The contradiction is not lost on me.

Having been processed I was free to find somewhere to rest my head. The town centred around a small port on the Rio Chucunaque, one of the many inland waterways in the area, and everyone congregated around a few noisy bars on its small main street. At each bar, reggaetón blared from large speakers and groups of inebriated men watched football or played cards. Chickens were grilled and beer bottles piled up outside. While this may not sound particularly appealing, there is something strangely inviting about it. As it was the final of the Spanish football league this was a special weekend. This, and the cockfighting tournament the next evening meant that Yaviza was going to be busy as people were coming from far and wide to enjoy the festivities. Many of them would be making their way in from the Darién jungle where they either farmed or logged, and these short visits seemed to include as much beer as possible.

By chance, I found a small hostel that had a ground-floor room that could accommodate my stroller. The building was on the side of a dirt road with the owners sitting under a house opposite. There were no check-ins or formalities; I just wandered up to the group of ladies and they barely acknowledged my presence. On asking about the room price, they muttered something acceptable, so I handed the older of the ladies the dollars in exchange for a key and a finger that indicated the hostel across the street. To say it was basic would be an understatement, but it was secure and had a bed and that was all I needed.

I had bought a few beers from the local mini-mart and remember sitting naked on my bed feeling deflated and alone. To say this was the very edge of civilisation was no exaggeration and I had nothing in common with those around me. While feeling sorry for myself, an American voice was discernible outside and I sprang to my feet, excited at the prospect of someone to share my Darién experience with. I opened the door to find a guy about the same age checking in. He looked equally relieved to see me and we made plans to meet up later.

Eric was an American historian who was writing a book on the history of the Pan-American Highway, and he was in Yaviza to interview a few people. After a few beers, we ventured out to enjoy a little Yaviza nightlife. It was becoming clear what a big social event the cockfighting tournament the next evening would be. Finding solace in each other's company, we drank into the small hours, devising ever more elaborate plans for forays the next morning.

Predictably, we both woke with sore heads but remained determined to turn our time in the Darién into an adventure. We agreed to meet after breakfast and make a plan. Eric arrived with a young chap we had met the night before who had agreed to be our guide. Communications were somewhat difficult but we managed to explain that we wanted to venture into the Darién to learn more about fighting cockerels and how they were raised and trained. He smiled and agreed although it soon became clear that there may have been a mix-up. Our guide thought we were actually going to buy a cockerel and enter it into the competition, which I will admit was probably drunkenly suggested the night before.

At the local port we bought tickets to El Real, a small riverside community. Our vessel for the day was to be a long dugout wooden boat wide enough for two people and long enough to fit about five rows. These shuttle boats were also the local delivery service to the jungle-locked communities, so ours carried a rocking chair and a refrigerator for delivery. Predictably, our first stop was the SENAFRONT office for permission to enter the Darién without all the paperwork needed for entry to the national park. With everything taken care of, we were soon speeding down the muddy brown river that was fringed by dense jungle on both banks. Every so often, minor breaks in the jungle revealed settlements nestling in clearings with small huts on stilts and dugout log boats tied to trees. We came across a couple of young girls paddling along the river and offered them a lift to the next junction. And so life went on the waterways of the Darién.

Our journey into the jungle took about two hours and of course our arrival was greeted by the SENAFRONT and yet another sentry post. Once passports were checked and bags approved, we made our way into El Real. For a town that is not connected to the world by anything but the river, it was remarkably well established. The streets were paved and Cable & Wireless phone boxes stood at street corners for the inhabitants of the small, colourful houses. Clearly, they had had a more prosperous past. The main industry in this area was forestry, or should I say, deforestation. Large, menacing machines lurked on the outskirts ready to wreak havoc. I am not sure there is anywhere in the world that you can truly escape football and here was no different. A local bar on the main street boasted that they were screening the upcoming Barcelona match.

We weren't here for entertainment; we were here for research. Word was that there was a guy here who could hypothetically hook us up with a cockerel, so we made it our quest to find his training camp. With the tournament set for later that evening, there was work to be done and we met a group of young men there. In their individual coops, the cockerels all looked in very good condition and in no way mistreated. Cockfighting is a big sport here and there is a lot of pride in the whole process. The owner of the establishment pulled out various specimens and showed us why they were deemed to be strong fighters. He explained that there was a lot of extra preparation before entering a tournament and that the bird then needed to be transported to Yaviza to compete. To ensure it was physically and mentally ready for the evening, it would be kept it in a cool dark space until the event started. Then there was the dress code. When we think of cockfighting, we might imagine razor blades attached to the feet or similar, but here that was not the case. There are, however, fighting tools to attach. Each cockerel is fitted with a calamus, or the hollow shaft of a feather, to its legs. We also quizzed our seller why he would sell us a good cockerel if he too was going to be entering the tournament? He only smiled.

Lunch in El Real was a fun experience. We wandered around the houses until we found one with a sign indicating it served food. We knocked on the door and were admitted into the home to be greeted by a young girl. She informed us that chicken, rice and potatoes were available and once we ordered she sat us down at the dining room table and disappeared into the kitchen.

Before returning to Yaviza, there was only one thing that Eric and I wanted to do and that was to be able to say we had swum in the Darién Gap. Back at the river, we found the most accessible point and stripped down to our swimming trunks. With the locals looking on, a few warned us that the river was very dangerous due to the strong currents. Determined to tick the box, I made my way down and jumped straight into the murky water. We fooled around, oblivious to the small sewer outlet just upstream of where we were. I was completely unaware of just how badly this act of bravado was going to affect me later down the line.

Back in Yaviza, we did a little tourist shopping and went in search of the beautiful skirts that the Emberá ladies wear. These brightly coloured items can be seen all over south Panama but we soon learnt that they are a relatively new fashion addition. The vendors informed us that the fabric was imported from China and bought wholesale from Panama City. Not only is this a cheap way to buy clothing, but the main reason they adopted these colourful skirts is that they believe the bright colours bring happiness into their lives.

The tournament was to be held in a large barn behind one of the more popular bars. The entrance was off a side street and once a small cover charge had been paid, we were shown into a large open arena. On the right, the ring and stands were sectioned off with wire. On the left was a large open area where everyone stood around making predictions for the night's entertainment. Cockerels were tied to the stands waiting patiently for their turn in the ring. While the gathering was male-dominated, it was by no means men only. Wives attended in support and were just as eager to share their views on which cockerel might be triumphant. Owners ranged from the local farmers to the chief of police and some had come from deep inside the jungle. In the run-up to the event, men could be seen arriving on buses and boats with small cockerel carrying cases under their arms.

A cockerel fight is essentially a wager between two contestants as to whose will be victorious. Once a wager has been set, the contest can begin. At that point, betting opens to the spectators, who wave money around and shout raucously, just like in the movies. Somehow through the mayhem, everything is coordinated, and the fight is ready. The handlers prepare their fighters at the side of the ring, strapping the additional claw to their feet. To add to its effectiveness, this is dipped in lemon juice to cause the other cockerel more aggravation if a strike is successful. The two birds are then introduced into the ring and circle each other, sizing each other up like prize fighters. They sway left and then right in search of any sign of weakness and then all of a sudden, with a flutter of feathers the fight is on. Each launches at the other with conviction, determined to vanquish it as quickly as possible. They tangle together and roll around on the dirt floor while the spectators rise to their feet and the noise escalates. Cheers and jeers are directed at the two birds as they dance, prance and strike. Feathers fly up and the proud handlers look on as their cockerels engage in combat.

Most people might assume that cockfights are fought to the death and while that might be the case in other countries, it certainly is not the case in Yaviza. Cockerels are expensive, and rather than see their property perish, the owners watch and evaluate whether their contestant has what it takes to win the bout. If a fight becomes one-sided, then the handler intervenes to forfeit his wager and rescue his cockerel. We stayed for about eight bouts and while the individual fights were aggressive, never was there a point that was repulsive to watch. The attraction for the spectators seems to be the betting and social occasion rather than any bloodshed.

My bus back to Panama City was scheduled to leave at 3am, so the night's festivities were cut short and I returned to the hotel to pack up and make my way to the terminal, where I had to oversee my stroller being precariously tied to the roof of a very overcrowded minibus.

Waiting to leave Panama, I reflected on the last nine months. I had covered 9,700km, run in eight different countries and literally worn through eight pairs of running shoes. In each country I had received nothing but support and at no time had I been put in difficult situations. Now I was preparing to fly back to Europe to celebrate my sister's wedding, but my mind was already focusing on having more than 7,600km still to run. The second part of my adventure would be where the real challenges would occur. It would take me through the Andes in Ecuador, across the Sechura and Atacama Deserts, before crossing the Andes once again from Chile into Argentina. I felt a mixture of excitement and sadness, not relief that I was going home. In this environment, I had thrived. The person I had become was a marked difference from the daydreamer I had been in London. This chapter of my expedition had highlighted the changes I wanted to make and South America would provide the venue to make them.

PART THREE - SOUTH AMERICA

CHAPTER 19 - FLYING BACK TO THE ADVENTURE

DISTANCE TO BUENOS AIRES: 7,633KM

My time in France had been highly enjoyable. While it hadn't been planned when I set out on this expedition, the chance to see my family and friends, even for a short time was refreshing. It allowed me to take stock and appreciate just how much I relished being alone on the open road.

The break from the adventure was not without its challenges, however. I found it hard to be around people for long periods of time and when the party atmosphere around the wedding reached fever pitch, I found myself disappearing to places where I could just sit and be by myself, unused to the constant stimulation of being surrounded by people.

Another hurdle was discovering that I had picked up a parasite known as Giardia in Panama, most likely from swimming in the Tuira River in the Darién. The symptoms were unpleasant and while I missed the constant momentum of the adventure, it was a relief to stop briefly to recover, as it probably would have rendered me useless in Colombia. Luckily, with easy access to medication in Europe, I was able to get this under control before the wedding celebrations commenced. The gift from Jen in Panama City that allowed me to bring my flight forward by ten days was even more fortuitous on reflection.

One of the strangest things that happened during my time away from the road was rekindling my relationship with my ex-girlfriend. We had broken up nearly a year before my expedition and when I left I honestly thought our differences would never be reconciled. But being alone on the road had given me time to think and over the course of numerous conversations we'd managed to get to a place where we wanted to see each other again. So, while on hiatus from my adventure, over a couple of days in Biarritz we rekindled our relationship. I have to take my hat off to her because visually and financially, I had changed a lot from the man she had first met and to add to that I was battling a horrendous parasite.

When the celebrations ended and the guests returned home, I was more than ready to escape back to the other side of the world. This interlude had been fantastic and my heart swelled to see my little sister so happy, but it really hit home just how much I craved solitude. More importantly, it reaffirmed my decision to pack in my old life in pursuit of something pure and simpler. So, on 1st July I was delivered to the airport by my older brother and his family and back to my life on the road.

Once on board the plane, the options for my approach to the second half of my expedition bounced around my brain. At the wedding, I had rashly set a potential end date of New Year's Eve, not based on clever calculations or anything like that, it was just a clean and tidy date to aim for that meant I could start 2016 with fresh plans.

The biggest change was going to be alcohol. Alcohol had played a big part in my decision to embark on this adventure and over the first half of this adventure, beer had increasingly wrestled its way into my routine. At first it was just a quick cold one, but the further I went the more frequent the mini celebrations had become and the quantity was increasing unnecessarily. Drinking per se wasn't the concern, I was more intrigued about how it might have been affecting my performance and enjoyment. Sitting on the plane, I took a sip from cold can of lager and resolved this would be my last beer for the duration of South America, not knowing that it would be a lot longer until I would enjoy another.

Once back in Panama City I headed straight to Luna's Castle to retrieve my stroller and all my kit. Knowing what happens in hostels, I was apprehensive about what condition my kit would be in or if it would be there at all, but to my surprise everything was present and correct. Having bounced, rolled and skidded nearly 10,000km over some pretty rugged terrain, my stroller was in need of some spare parts, which I had ordered from Thule before leaving for the wedding. Facundo, a colleague of my sister-in-law, had very kindly agreed to take delivery of them and duly delivered them to my hostel. With my stroller back to full strength, I started to plan my crossing to Colombia.

As with the crossing from La Paz to the Mexican mainland, a route less travelled was calling me. Research unveiled an option to hitchhike along the east coast of Panama, cross the border to Colombian waters and arrive in Turbo. Without delay, at 6am I strapped my stroller to the top of a truck while chatting to a bunch of Irish backpackers who were returning drunk from the local brothel, clearly baffled at this harebrained scheme.

After a three-hour bumpy ride, I arrived at the Puerto de Carti and set about finding a boat to transport me to the Islas Carti, a few kilometres offshore. The port was nothing more than a shack on the shore with a jetty protruding into the Caribbean Sea. Litter washed up, while crates of beer and Coca-Cola were readied for transportation in long, thin boats with thin awnings to fend off the oppressive sun and ocean spray. Most of the passengers were tourists on their way to the San Blas Islands, with the remaining handful there for onward travel. With my heart in my mouth, my running stroller bounced around on the bow of the small boat, treacherously close to becoming dislodged into the sea. Perhaps this adrenalin helped, because a sense of adventure was flooding back, and I was chomping at the bit to get started.

The Islas Carti filled me with equal measures of disbelief and wonderment. Where I landed, the island was only a few hundred metres long and half as wide, however, not even a scrap of land was visible as small huts clung hazardously to every available inch. Each shoreside shack had a makeshift jetty that supported a small, square room built of corrugated iron. I would later learn these were the toilets. Between properties, the odd coconut palm would squeeze up where possible and reed barriers provided some privacy and protection from the wind. Our small boat pulled up alongside my hostel and we carefully lifted my stroller onto the thin wooden jetty. Brightly coloured paint gave the property a certain charm but could not detract from the hundreds of plastic bottles and other garbage that bobbed between the stilts and onto the shore. My stroller had to stay in the open bar area while I would sleep in a rustic bunk room upstairs. This was my base until I could find transport to Colombia.

Without any regular service or timetable, this was a case of finding out which boats were heading south and securing a place. My fixer would be a local chap called Negra. He assured me that he would be able to arrange safe passage and advised me just to wait and be ready. Even at this stage, it was becoming clear that I may not have brought enough cash for this particular excursion and that this route might not have been the ideal choice for someone with a deadline.

With little else to do, I set about exploring the island. My hostel comprised a main bar area, a jetty with a toilet that dropped directly into the sea, a kitchen with pizza ovens and a small outside area for food preparation and cleaning. From here, a steep staircase led up to the dorm rooms above.

Outside the property, dogs sniffed around the warren of narrow streets where children did their homework on street corners. Doors to many of the homes hung open, giving a glimpse of the tight living conditions inside. A wider street ran up the spine of the island, where a few colourful stone buildings served as restaurants and mini-marts. Old Cable & Wireless phone boxes mourned their previous usefulness while shops stocked only the bare essentials that wouldn't perish. On the other side of the island, fishing boats were prepped near a more established pier that offered views of several other islands, all brimming with huts as the sun set behind them.

On my second day on the island, I pushed Negra for an update on progress and he managed to evade my enquiries without providing any concrete information. Another would-be passenger was also getting visibly irate. The other islanders I tried asking were equally quixotic. I decided to give it one more night. Boats came and went and the restaurant continued to feed me but I was growing restless.

On 5th July I woke up to find out that a boat doing the exact journey I wanted to do had been and gone during the night and my fixer, Negra, had missed the opportunity. Another boat might not come for a few days, so seething with anger I went in search of answers. Even on this tiny island, he had disappeared, which was doubly annoying. Surrendering to impulse, I stormed back to my hostel and demanded a boat back to the mainland. Unprepared to waste valuable time and with no assurance of onward travel, I had no option but to regroup in Panama City and find an alternative route. As my boat puttered away from the island, Negra appeared on the end of a jetty, clearly annoyed that my departure stopped him earning any more money from me.

Back at Luna's Castle for the fourth time and with my tail firmly between my legs, I settled for a plane. A backpacker was also looking to get to Colombia by a less established route and between us, we found flights that would take us to Puerto Obaldía on the Colombian border.

Our flight skirted the coast of Panama, where nothing but dense forest lined the coast and waves pummelled the rock. The occasional small fishing village nestled in a cove and I'm certain that during the short flight I saw the islands I had been stuck on.

Puerto Obaldía was a small fishing village only accessible via plane or boat, the nearest road being the one I had run to Yaviza a month before and nearly 100km away. Our small propeller plane carrying just five passengers touched down on the runway, after a descent that had me convinced we were going to ditch into the ocean.

To check out of Panama and gain entry to Colombia meant being ushered through the village to a military post. My last experience on Panamanian soil was the ransacking of my stroller by a soldier before he would allow me onto the jetty to board the boat to Capurgana. Frustrated at having to repack everything, I conceded that this route was extensively used to traffic drugs and he was only doing his job, though he didn't have to look so happy about it.

Excitement infused my body as our boat bounced its way along the Colombian coast. With every hour that passed I was getting closer to reclaiming the road and forging my way through South America. It felt like being back where I belonged.

After stopping for fuel at Sapzurro, we finally arrived in Capurgana where I placed a foot on Colombian soil. Despite this town's remoteness, or perhaps because of it, this seemed to be a tourist destination, mostly for Colombians. Buildings lined the seafront, offering tours to the San Blas islands and beyond, while fresh fish and cold beer could be had from the multi-coloured restaurants and bars that perched on the rocky coast. Behind the main waterfront, the streets were more rustic but maintained the same casual vibe. Colourful graffiti was daubed on rundown buildings between pristine churches and dark wooden houses, and if you wanted to get around, a good option would have been one of the horse-drawn carriages with makeshift plastic seating. This utilitarian efficiency meant that the airport next to the main square also doubled up as the town's football pitch.

Revelling in forward momentum once more, that night I settled down early, conscious that an imminent boat crossing to Turbo would finally have me back on the road again.

CHAPTER 20 - A MARATHON A DAY FOR 178 DAYS

DISTANCE TO BUENOS AIRES: 7,633KM

A marathon a day for 178 days without rest was the challenge I had inadvertently set myself at my sister's wedding. On making the bold assertion that I would arrive in Buenos Aires in time for New Year's Eve 2016, I didn't know the real distance from Turbo in northern Colombia to Buenos Aires. My focus on arriving in Panama on time had been so rigid that I hadn't given much thought to what lay further south. Averaging just over 30km per day in North and Central America had included a lot of rest days and logistical interruptions. As I sat on the motorboat gliding over the Atrato River to Turbo, I had my first doubts about whether I would even make it.

This segment of my journey felt strikingly different from my time in North and Central America. Now, I faced a pressing deadline—an ambitious challenge that demanded my full commitment. To succeed, I would have to immerse myself in the relentless rhythm of running, eating and sleeping, leaving less room for enriching interactions with the world around me.

This shift didn't diminish my enjoyment; rather, it raised the stakes. My sense of purpose became urgent as I prepared for the next 178 days, aiming to average a marathon a day. Even now, I wince at the formidable challenges that lay before me. It wasn't just the distance that made this journey daunting; it was the prospect of navigating the unforgiving landscape alone that truly drove home the gravity of my undertaking.

In contrast to the first two parts of my adventure, where exploration and discovery reigned, this phase was a race against time. Long stretches of my days became a steady cycle of routine, reflecting the intensity of my focus. Parts one and two were journeys of discovery; part three was a relentless pursuit that would push me to my limits.

Turbo itself was a stark contrast to the calm crossing. At the entrance to the wharf, buildings balanced on wooden platforms and locals clambered over colourful boats that were lashed together to sell their goods and services to all who disembarked. I could sense this wasn't a place to hang out and exited the port as quickly as possible to arrange everything for the long day ahead.

After a quick restock, I was finally back on the road and after a month without running it felt amazing to be back in the environment where I thrived. South America was the part of the adventure that held most excitement for me. Not only did it seem as if the Darién Gap and lack of a connecting road had slowed the influence of the USA, but this leg of my journey was also where I was going to meet my biggest challenges. In just a few days, I would ascend to my first proper altitudes, and further down the route, I would have to take on the Sechura Desert, the Atacama and the Andes Mountains. Things were about to get real.

As it had been a long time since I had been on the road, I tried to keep it short and set my sights on the small town of Apartadó, some 31km away.

Colombia had a distinctly different feel to the other countries I had traversed. The Caribbean influence was far more apparent in the colours of the buildings, the food as well as the energetic music jumping from every shop, restaurant and passing car. Despite all the warnings I had received, this land felt welcoming and eased my transition back to life on the road.

So, it was frustrating that the next morning my plantar fascia was making its presence felt once again. The first time was at the beginning of my adventure and then again after my break in La Paz.

In the knowledge that the cause was a sudden increase in mileage, I rued not continuing to run while in France, even though my excuse was Giardia. On the positive side, I had overcome the pain in my foot before and knew how to deal with it. But that was no comfort on the 25km of straight road to Chigorodó because it already felt like reaching Buenos Aires on time was going to be unlikely.

Now the normal way to deal with an injury would be to stop, rest and recover but that isn't how my brain is wired. With somewhere to be and a target to hit, pain was not going to get in my way. Mutatá was 57km away and I knew I was going to need distractions to help me get there, despite recently having embraced tunnel vision. Sometimes success is more about adapting along the way, and this was one of those occasions.

About halfway through the day I ran past a school and not long after was tracked down by a young teacher on a motorbike. He had spotted my stroller and thought it would be interesting for his pupils to learn about what I was up to. With a little persuasion, he convinced me to turn around and go back. I was soon surrounded by a group of teenage kids all talking at once. They didn't seem very interested in my journey but more what music I was listening to. They shared their reggaetón music, a style that originated in Puerto Rico during the mid-1990s and has been influenced by American hip hop, Latin American, and Caribbean music, and in turn, I shared some Scottish Pipe Band music, much to their amusement. They managed to persuade me to do the highland fling but seemed reluctant to share their dance moves.

Later that day, a local carpenter approached me and invited me across to his workshop. He spoke quickly and enthusiastically and to be honest I didn't really understand what he was saying. As I nodded along politely, he picked up a thin plank of wood with a long text delicately engraved on it and insisted I take it. I'm not going to deny that I was less than enthusiastic about carrying another piece of wood and this time for over 7,600km to Buenos Aires, on a stroller that was already overweight.

After a few days of fairly flat terrain, the road and the scenery started to transform. Banana plantations were replaced with lush mountains and roads that endlessly rose and fell. An email from my father that day informed me that I had a mostly flat day before the pretty serious hills started. After hours of pushing the stroller up hills and it pulling me down again, I questioned my father's description of mostly flat. His answer was short: 'It is flat compared to what is coming tomorrow.' I felt the dread seeping in.

It was in Mutatá that I had my first run-in with the Colombian military. Early that morning I was just getting ready for a long push to Dabeiba, some 55km further down the road. I had stopped at a stall to buy breakfast from an impeccably dressed little old lady who must have been over 70. She had a barbecue fashioned out of an old barrel, a plastic table and a worktop all sheltered by the thick foliage of a tree. As she prepared me a strong coffee, a soldier in full battle dress approached with a machine gun strapped across his chest. No matter how seasoned I felt by this point, there is always something intimidating about the military, especially in this part of the world. Here they know that they can stop anyone and ask anything. I braced myself but was relieved to be greeted with questions of interest rather than interrogation. Our conversation started to draw an audience as his comrades in arms gathered around, displaying disbelief and admiration in equal measure. As talk petered out, I pushed on, with the soldiers' good wishes propelling me forward.

Cañasgordas is a relatively large village nestled high in the hills of Antioquia. At 1,288m, it marked the highest point of this expedition so far, and the altitude meant the temperature cooled, which was a relief. However, my plantar fasciitis was really aggravated with the change in gradient and after having run 200km since arriving in South America, I was forced to take a day's rest. My mileage had been dropping and the pain was becoming unbearable. Despite being well below the daily average and needing to get to Buenos Aires without a day of rest, I was risking more serious injury. To make matters worse, the Giardia I had picked up in the Darién region of Panama was flaring up again. The day off was spent at the pharmacy stocking up on medicine and painkillers, which thankfully came in much stronger doses than back home.

Having a day off afforded me a little time to soak up some of the local sights. With a spectacular valley setting, the main tree-lined square of Cañasgordas is dominated by a large white church with two spires. Small shops crowd the streets and it seemed very peaceful and inviting against the green hillsides. However, a darker side still seemed to lurk in this part of Colombia. Just a month before my arrival, on the road I had just run, armed guerrillas had set up a fake checkpoint and murdered three people. I had to remind myself to keep vigilant and not take unnecessary risks.

CHAPTER 21 - I START TO CLIMB

DISTANCE TO BUENOS AIRES: 7,431KM

The rest had done the world of good and over the next four days, I covered nearly 190km on my way to La Pintada, bypassing Medellín, notorious for its connection with Pablo Escobar. It seemed sensible to try to skip the larger towns in Colombia because it would be a lot safer, but also because I preferred the less explored parts of the country.

The route from Cañasgordas to Santa Fe de Antioquia took me on a staggeringly beautiful 200m ascent followed by a long, winding descent with spectacular views across the valleys below. Santa Fe is one of Colombia's national monuments, and its beautiful, cobbled streets, colonial buildings, impressive cathedral and elaborate suspension bridge over the Cauca River had attracted a collection of mostly local tourists.

From there, I followed the main road until a small detour to Anzá. It was a little before where I wanted to stop for the evening, but with my recent niggles I decided to pause there. The route wound along a quiet road that loosely followed the brown river at the bottom of the valley. Thickly wooded hills rose up on either side and there was very little in the way of traffic before I stumbled on this quaint village built around a small tree-lined square. At one end, pigeons roosted on a beautiful stone church with white-painted pillars and ornate ledges, while the other sides were fringed with small tuk-tuks and a couple of restaurants with colourful plastic furniture and sunshades advertising local beers. Older people sat on benches as the children played. Some places just stick with you, and this is one of them for me. The memory of sitting on a doorstep the next morning, nursing a small plastic cup of strong Colombian coffee, is as sharp as a photograph in my head. On the descent down the short hill to the road below, mist washed over the valley floor and a beautiful white horse stood in the middle of the road, blocking my way. The whole experience seemed so mystical, it set me up perfectly for the route to Bolombolo.

Bolombolo is another small town on the banks of the River Cauca, just off the main route. Here small colourful houses lined the streets, while fresh fruit and vegetables for sale overflow from plastic baskets. In a basic room in a small hostel above a saddlery, I settled down for the night, knowing another big day lay ahead of me. Getting to La Pintada was not a done deal. As the night drew in, it became apparent that blissful rest would elude me as the small restaurant across the street had reinvented itself into a karaoke bar and drunken renditions of Latin ballads shattered the tranquillity.

The challenge of making it to Buenos Aires on time weighed on my mind. After only 11 days in Colombia, I was already a full day behind schedule. My only chance of getting there even remotely on time would be to increase my daily distance.

Leaving La Pintada behind was imbued with an urgency to recover lost ground. Food, however, was harder to find. At one point a local chap insisted that I accompany him to his home, which was not much more than a rundown shack on the side of the road. The more insistent he became, the more uncomfortable I felt and made my excuses to push on, claiming I had somewhere to be. There were many occasions on which I perhaps didn't fully take advantage of the hospitality on offer, but in this instance, I was confident I made the right call. Far better to wait for the little village of La Felisa, which was nothing more than a few shops and restaurants lining the road. At least there I could safely take a small room above one of the confectionery shops that sold large chunks of processed sugarcane.

My next challenge would be to get to Cerritos, where a friend of a friend had very kindly arranged a place for me to stay. With 110km to cover of this country, I was determined to sustain my new surge and get there in two days. The road seemed to climb continuously, and the heat of the sun was relentless. Having somewhere to stay and the prospect of some home comforts was all that could maintain my focus, helped by minor distractions of the road. For example, I got talking with a man on a huge BMW bike while stuck in a traffic jam before a toll booth. He was keen on adventuring, but the similarity ended there. His immaculate new motorbike with matching bags and well-thought-out sense of style only highlighted just how tramp-like my appearance had become. So much so, that after realising how I appeared, I emailed my imminent hosts to assure them that under all the hair and grime, there was a decent bloke.

My stay with Tony and Lizzy started with a real welcome. They treated me like a king from the moment I arrived and knew exactly how to comfort someone who was physically and mentally drained. After days of cheap makeshift hotels, cramped bathrooms and very little luxury, I relished being able to let my guard down for a night and properly recharge my batteries. When you are constantly moving and trying to sort logistics while worrying about safety, your brain can never really stop, relax or recover. In a gated property, however, where everything is taken care of, the relief is overwhelming.

Despite wanting to take Tony and Lizzy's kind offer of a day's rest at their beautiful house, I had to push on. At less than 600km in 15 days, that marathon-a-day average was forever front of mind.

Colombia's developed areas tended to be concentrated along corridors, and the roads heading south through La Victoria, Bugalagrande and onto Buga were nice and wide, even if full of heavily laden lorries screaming past.

Buga dates back to 1555, making it one of the oldest towns in Colombia and home to the Basilica Menor del Señor de Los Milagros, or Lord of the Miracles. It is believed that the dark depiction of Christ in this pink church came into existence miraculously, or in other words spontaneously and without the work of human hands, making it an acheiropoieta. Since then it has become a destination for huge numbers of pilgrims in search of a miracle. When I paid a visit, the church was teeming with people, some of the more than 3 million people said to come each year, over three times the number of visitors to Canterbury Cathedral. While I am not religious, there is something alluring about so many people gathered together to pray for something positive to happen in their lives. It didn't escape me that this was very similar to my motivation for running, that is, to find some sort of meaning and purpose to my life.

Another point of interest in Buga is an inland lighthouse in the southern part of the town. While the crowds flocked to see the lord of miracles, this tower was deserted apart from the two security guards who let me climb the internal staircase. While not viewed as a miracle like the nearby carving of Christ, for me this tower held a very special moment as it was one of the rare moments that I was high enough up to see both the route I had travelled and what lay ahead. Having this peaceful pause allowed me time to reflect and check in with myself.

My routine was changing as I grew stronger and my injuries had subsided. Feeling better had unlocked the early mornings, allowing me sometimes to complete my distance by lunchtime, giving me time to explore, recover and soak in the abundance of culture the local area had to offer.

The border with Ecuador was now just two weeks away. Sugarcane plantations provided the backdrop as I ran through the Cauca Valley, dodging the huge harvesters and tractors that roamed these roads. My aim was the small town of La Bolsa just south of Palmira, and on the way there I had a few conversations with people on the road. It turned out that La Bolsa wasn't the kind of place where a gringo would be that safe. The harvest attracted workers from all over Colombia and this sometimes led to an uneasy atmosphere involving people drinking and looking for trouble. I was strongly advised to continue a few more kilometres to the town of Villa Rica.

Villa Rica proved an unexpected pleasure and remains one of my favourite overnight stays. My hotel would be a small, basic establishment on the main square, served by a small ice cream vendor directly outside who sold the largest ice creams I have ever seen. For a small sum a plate was set down, overflowing with ice cream, exotic fresh fruits and whipped cream, slathered in caramel sauce. As I battled this sugary mess, stallholders, horse-drawn carts and locals in brightly coloured clothes all got on with their lives around the square. This included the meat vendor, whose large joints of meat swayed from hooks, all shielded by a shade stretched over bamboo stakes. Ingeniously, his chopping board was a large tree trunk fashioned into a table, with dogs lurking expectantly around a bucket full of cow hooves. As the day turned into evening, the stalls were packed away and couples took to the streets to literally dance the night away. This moonlit tranquillity allowed my brain to switch off and appreciate the Colombia I had wanted to experience.

It was market day when I arrived in Santander de Quilichao and the streets were alive with people selling fruit, vegetables, sugar products and meat. Huge tarpaulins hung from building to building to shelter the produce from the heat of the day. Ladies, dressed in traditional top hats, wandered back and forth spraying water to stop the vegetables from wilting, mingling with workers, horse-drawn carts and homeless people as they went. As the day wore on, the smell emanating from the market became sweetly repugnant as huge piles of waste began rotting in the heat. Teams of sweepers appeared from nowhere to attack the mess, causing dust to rise and filling the streets with a dense haze. A brightly coloured chicken bus parked in the middle of the street caused chaos while the goods bought or still unsold were loaded onto the roof. Children played in the streets, ducking and weaving between the piles of rubbish, while workers shared jokes and those who had drunk too much slept it off in doorways. Huge numbers of vultures glowered on the terracotta roof tiles and opportunistically swooped down for morsels.

South of Santander, hills swelled into mountains, and I began to feel the added effort that required. The stroller felt ever more leaden as I coaxed it up narrow winding roads, battling the oncoming lorries, humidity and heat. Having run 850km, I was two-thirds of my way through Colombia and while I loved being here, I began to feel the pull of Ecuador starting to draw me in, despite my body starting to creak again. Running for 18 out of the last 20 days necessitated a rest, so at the beautiful town of Popayán, also known as the White City, I hunkered down for a couple of days in a small hostel to regroup. The first person I met on entering this city was an old American chap on a bicycle. He was unshaven, wore a faded cap with a striped tee-shirt and while seemingly unapproachable, once again first impressions were so wrong. The most memorable thing was how concerned he was about my safety and, more specifically, my lack of eye protection, due to having lost my glasses somewhere along the route. He immediately took off his fake Oakley sunglasses and gave them to me, not taking no for an answer. Taking me under his wing, he rode alongside for a few minutes and directed me to where I needed to be.

Popayán is a fabulous place to unwind. The town's jaw-dropping beauty is enhanced by its wide streets of gleaming white buildings, set off by terracotta roof tiles and ornate doors in an array of striking colours. Dominating the main square is the Cathedral Basilica of Our Lady of the Assumption, with its impressive neo-classical columns and statues preaching from the roof. The verdant square below bursts with trees and shrubs in pristine condition. Amid the buzz of other people's daily routines, it was great to just relax after days on the road. But I soon learnt not to relax too much. While out collecting supplies for the next surge towards the border, I was almost victim to a robbery. Engrossed in my phone to navigate the city centre, I didn't see the bike fly down the street with a kid hanging precariously onto his friend. With hand outstretched, he tried for my phone but luckily his aim was off and nothing more than a brief collision occurred. When alone on the open road, you are constantly on high alert. However, in a city if you let that guard down for just a few minutes, you are easy prey. I wasn't angry at the kids as this thing happens all over the world, but I was frustrated at myself for my moment of weakness.

The roads south of Popayán were not only hilly but increasingly dangerous. Tight winding roads without hard shoulders and numerous blind spots provided plenty of hairy moments as huge lorries came roaring up behind me. Despite the near-death moments, the adventure was still providing plenty to keep it rewarding. Sometimes all you need to take your mind off the danger is a car to pull over, proffering some money and a nice cold can of Coca-Cola.

One of the highlights was an unscheduled stop at a small roadside panela factory, or more accurately, a small wooden structure with a corrugated iron roof and rough stoves in the middle. Panela is unrefined whole cane sugar in solid form, derived from boiling and evaporating sugarcane juice. On one side of the hut was a pile of raw sugarcane and on the other, a pile of crushed husks that had been processed through a rusty looking diesel-powered pressing device. The pure sugarcane juice is collected in a barrel topped with a thick foam, and from there it is taken to three large boiling pans to be reduced down to a thick caramel-like syrup. This is then poured into wooden moulds and left to harden into panela and offered for sale at numerous stalls along the road. The team of friendly workers welcomed me in and took me through the process, allowing me to taste the panela at each stage of its production. It may have been a short interaction, but it remains a highlight of the journey.

El Bordo also stuck with me, but for different reasons. Waking in a dingy hostel, I sat eating my granola with water from the tap in the bathroom. The weather was perfect and the road, while not flat, was smooth and inviting. Around every corner a glorious view awaited down long valleys painted a full spectrum of greens and browns, punctuated with the occasional bright purple jacaranda or yellow guayacan flowering tree. Every time I needed food or water, a village seemed magically to appear. My video footage of this episode contains gripes about how hard it is to get up in the morning but how rewarding it is once let loose into the mountains. It finished with the proclamation, 'I love my life!' which still brings a smile to my face.

What actually makes a day like this so memorable is the unforeseen challenges that crop up. Having completed over 40km, I arrived at my target hostel to find nothing but a fenced-off and boarded-up shell. Barbed wire lined the roads, precluding any discreet wild camping, so on I battled through the 50km mark and up into the mountains. As I raced the setting sun, a resident of a small cluster of shacks greeted me. On hearing my predicament, he dashed off in search of a solution. Minutes later, we were walking to a disused restaurant where there was a spot I could use next to a pool table. There I was, dirty, tired, and hungry in an open-sided building and all I could think was how lucky I was to be living this life.

Long sections of road construction made the roads pretty treacherous for two whole days. Infrastructure unravelled as tarmac disappeared, any excuse for a hard shoulder vanished and heavy traffic was funnelled into a tighter space. My lungs were scorched with dust from the passing lorries and my eyes clogged up with dirt. My only hope was to focus on the border ahead and the next chapter of my adventure.

Hitting Ipiales came with a massive sense of relief and achievement. I had been in Colombia for exactly a month, and had covered 1,270km with only two days' rest in that time. While a fraction behind schedule, I was feeling strong and more importantly, could look back on my time here with only the best memories. The Colombian people had been so hospitable and their country left a lasting impression. It also introduced mountains, the like of which I had never encountered, so the running was getting harder, but these gradients were nothing compared with what Ecuador, Peru, Chile and Argentina had in store.

While in Ipiales I had some time to kill due to an interview with the Marathon Talk podcast, so added a visit to Las Lajas Sanctuary, a towering Gothic-style Catholic church on the arched bridge that spans the Guáitara River in a steep-sided gorge southeast of town. To get there I shared a taxi with people who were on their own pilgrimage to visit this spectacular site. I donned my tourist hat and allowed myself to forget the road, just for a few hours.

CHAPTER 22 - BRUTAL RUNNING

DISTANCE TO BUENOS AIRES: 6,396KM

After the usual border-crossing palaver, I was soon released into Ecuador, the eleventh country of my solo run from Canada to Argentina. With 10,614km behind me, the remaining 6,350km weighed on my mind and if I was going to make Buenos Aires anywhere near my self-imposed cut-off of 31st December, then I would have to keep moving. With 147 days to go, my required daily average had crept up to over 43km a day.

My introduction to Ecuador was brutal, and the altitude was already 2,800m above sea level. That first day would see me ascending to over 3,300m. Running at altitude was unavoidable and the naysayers' voices span around my head, saying that I wouldn't be able to make it. Battling the scepticism they had sown caused me some conflict, as the necessity of securing over a marathon a day was very real. A flash of self-doubt crossed my mind, as is only normal when presented with a challenge of this scale. However, I had also learnt not to let worries like this affect my progress. The trick was to reach back and remember all the things I had overcome already and be very realistic about the much larger challenges ahead.

As this internal debate raged, I was suddenly sent spinning off the road after something hit me on the left elbow. In shock, I swung round to see a small motorcycle speeding over the crest of a hill. Fortunately, the pain was not too great as I took stock. After over nearly a year, this had been my closest encounter with the many vehicles I shared the road with. While I was relieved, it also struck home just how lucky I had been. If it had been a car or even worse a lorry, then my story could have been very different. In this case, it was just a box strapped on the back of the motorcycle that had grazed me and within a few moments I was ready to continue.

As is often the case after a sudden shock, the doubts and overthinking that had immediately preceded the minor collision all but vanished, to be replaced with adrenalin and renewed determination. This was a new country with new challenges, and I wasn't going to back down. That night I arrived in the small city of San Gabriel, one of five Pueblos Mágicos or magical towns, in Ecuador. However, the streets were subdued and very little was open, leaving me to resort to just a sausage and a potato on a stick I bought from a street vendor. While this meal contained no real nutritional value, at this hour and with tiredness in my bones, the warm fats made it delicious.

Quito, the capital of Ecuador, was just four days away and I had to think about logistics and tactics. I had contacted the Adidas PR team, who had very kindly sent nine pairs of shoes to France to help me complete the South American stage of my run; the only issue was that I couldn't bring them all with me. The solution was that I brought over two pairs and my mother would send two pairs to Quito and then two pairs to Lima so I could replace my running shoes before I sustained any more injuries. In Quito, we had chosen a small hostel that had kindly agreed to receive the package for me.

This was timely because my left foot was starting to flare up again with the long steep descents of the Andes Mountains. On one hand I was cursing my own weakness, while inwardly chanting that the pain was necessary to experience all the beauty around me. And at that moment, it really didn't matter how painful it was or that I was standing on the hard shoulder of a busy highway because my eyes were feasting on the rugged mountains stretching south, getting ever bigger and more imposing. These mountains were not an impassable obstacle but a beautiful challenge.

After a night in Ambuqui, I set my sights on the town of Ibarra, about 40km south along the main highway. For the first half of the day, the road followed the river Chota along the valley floor, with slopes rising on both sides of the road. The highway itself was clearly going to be expanded into a full-scale motorway and although it wasn't pretty, the surrounding scenery spurred me on. After about 15km, the road started to climb and for the rest of the morning, I pushed my stroller uphill and into Ibarra for lunch. Feeling strong, I decided to up the mileage and do the extra 25km to Otavalo, which hosts one of the largest and most famous indigenous markets in South America.

The main highway was becoming oppressive with the constant roadworks and heavily laden trucks and lorries, so I was relieved when an alternative route presented itself. While the new road was narrower with very little hard shoulder and the traffic still steady, it was definitely safer than the main highway. Small, run-down houses lined the road, in which new mixed in with old, and white graffiti adorned many of the walls.

As the kilometres ticked up and the altitude started to bite, I could feel myself having to dig deeper. The stupidity of my decision to push on to Otavalo didn't escape me. In all honesty my new plan was not about experiencing the famous market or exploring the streets, it was simply an excuse to keep pushing south, to add valuable kilometres. But it was also true that despite hating big cities, Quito was drawing me in like a moth to a lamp. By the time I rolled into Otavalo, I had covered over 60km and climbed over 1,200m, all while pushing a 40kg stroller.

After securing a place to sleep, in the darkness I went in search, not of the famous sites or beautiful buildings I should have visited, but for a sim card and supplies. But more immediate priorities took over when I suddenly came across faint and had to seek support from the nearest wall. Not knowing what was wrong, all I could think about was refuelling and I staggered into a small burger bar and ordered a beef burger, fries and a large Coca-Cola. When I had devoured that, I ordered a chicken burger, fries and a large Gatorade. Bizarrely, while my brain should have been thinking about sensible steps to recover, it was once again only focused on the steps to cover and more importantly, the 100km to Quito, where I would be picking up a package from my mother.

The area just north of Quito is undeniably beautiful, with mountains and volcanoes standing proud. To my west, the snow-capped Volcan Cayambe soared to 5,790m and unbeknownst to me at the time, I would be back in a few years to stand on its summit. For the first section of the day, the main road bisected farmlands with tractors, lorries and workers all hard at work. Today's bonus was that the route would reverse the 1,200m climb of the previous day and while running downhill is exhausting, it's more gratifying than slowly labouring up.

The other benefit was being able to witness more of the real Ecuador, in particular the people who lived here. The indigenous inhabitants of Ecuador were here long before the Spanish conquest, having arrived from Asia via the Bering Land Bridge. Nowadays about 70 percent of the population is of mixed indigenous and European heritage, but the remaining indigenous population is still in evidence. Women in particular are easy to spot from their traditional clothing. Typically, this includes a white blouse, a blue or black skirt, a shawl and jewellery that holds great cultural significance. These are topped with a fedora hat that sits on their beautiful dark hair.

In my mind, crossing the Equator and entering the southern hemisphere for the first time was going to be a big deal, but in reality it was something of a disappointment. The invisible line that divides our planet north to south crossed my path on a rather steep uphill climb without even a signpost to mark it. While it had no real bearing on my trip, to have run here from Canada was a moment of personal achievement. The adventure had taught me that I needed to celebrate milestones such as these, even if just to provide the energy to keep going for the rest of the day. It's probably quite telling that I don't even remember crossing the Tropic of Cancer when I was in Mexico as there was so much going on and I was still acclimatising at that point.

The final run into Quito was marked by searing pain and this would be in contention as one of my worst days of running. Although only 40km from where I had stayed, the stretch to the centre of town was a tough 1,000m climb along busy highways, accompanied by noisy lorries blasting fumes from their exhausts. These roads cut through large rocks, and in the heat and smog I fought choking claustrophobia in the never-ending outskirts of Quito. More and more I was becoming a point of interest and there was no avoiding a more threatening atmosphere here. The traffic became frenetic and I could feel tension rising inside me, just as my energy was draining away. Over the last seven days, I had covered 350km and climbed over 4,000m, pushing a stroller that fought as much on the descents as it struggled on the climbs. While battling on was mentally, physically and emotionally difficult, I couldn't ignore the fact that I was also very excited to be in Quito for a day off. But before I could enjoy any of that, I had to keep my head down and push to the end, knowing that the feeling of relief would make all the effort worthwhile.

CHAPTER 23 - STUCK

DISTANCE TO BUENOS AIRES: 6,133KM

Terrible news. The front desk of the hostel had not received the package I was expecting. This was tricky because any more delays would seriously impede my progress and there simply wasn't any room in the schedule for that. There was nothing else for it but to visit the local post office. Even with my pidgin Spanish, I could tell that the post office didn't share this view, and I was swiftly redirected to the local depot. The update I received there was less than encouraging. Having left France some 8-10 days earlier, the package had duly arrived in Ecuador but was being held due to a customs issue. It felt like I was being held hostage in Quito by the Ecuadorian postal service and that every moment I spent here was going to have a knock-on effect further down the road.

After just one day I was longing for the open road and to leave the metropolis. But returning to the post office made clear that the issue was completely out of my control, and I could be stuck here for an indeterminate amount of time. To remain positive I visited some of the tourist attractions and indeed enjoyed the freedom of wandering the streets without a stroller. These included the Plaza Grande, the impressive Basilica del Voto Nacional, the Convent and Monastery San Francisco, the opulent Jesuit Church of La Compañía de Jesús, and the home and gallery of Oswaldo Guayasamín. The art of Guayasamín, a prominent Ecuadorian painter and sculptor renowned for his powerful and emotive creations, really made an impact on me. A recurrent motif in his work is the human figure, which he portrays in various states of anguish and resilience and in that I found a strange connection. Framed posters of his art hang on the walls of my home today and every time I look at them I am transported back to South America and a flicker of the emotions I felt at that time.

During my stay, Quito was undergoing a turbulent time politically. Thousands of people were protesting proposed constitutional amendments, the expansion of the oil frontier, mining projects, changes to water and education policy, labour laws and pensions, a proposed 'Free Trade Agreement' with the European Union, and increasing repression of freedom of speech, among other things. As a result, hundreds of police officers had been drafted in from all over Ecuador to handle the nightly riots in the streets and to control the crowds. However, it was a different story in the daytime, when they seemed to have the same idea as me and could be seen around the various tourist attractions.

In Quito, I was able to eat like a king, no longer restricted to what could be found at roadside restaurants. One of my favourites was a little Italian pizza restaurant nestled on a side street in the older part of town. Here they made the greatest calzone pizzas and each night, under the guise of refuelling for the journey ahead, I could be seen back at the same table enjoying the same meal, time and time again.

By my third day in Quito, I was getting restless and the post office was offering very little hope. There were three options. The first was to sit and wait for my package to arrive but the thought of being held hostage was too much to bear and enough to immediately disqualify this one. Second was buying two new pairs of running shoes that would get me to Lima, where my next resupply should be waiting, but this would strain my already stretched finances further. Not only this but my mother had taken the time and expense to send the package here and it wasn't just running shoes inside. The final option was to continue my run as planned without the package and then return by bus to collect it when it was finally released. This would at least limit my wasted days to just two and that was all the persuading I needed. I started to plan my return to the Pan-American the next day, still clinging to the hope I could be in Buenos Aires for New Year's Eve and ignoring the fact that the daily average needed was now 44.5km per day.

It was on the evening of 14th August that the volcano erupted. Cotopaxi is an active strato-volcano 50km south of Quito. I can remember a large bang that evening and the excitement and fear of those on the streets. Even at a safe distance from the eruption, ash could be seen coating the cars in the streets of Quito. I also recall the mix of frustration and disappointment I felt, not because this was going to affect my progress in any way, though I was now advised to wear a mask when running, it was because the unplanned additional two days in Quito meant that I wasn't standing at the bottom of the volcano when it had erupted. Of all the things to experience in life, witnessing a volcanic eruption of this size would have been on my list.

Despite my frustration over time wasted in Quito, it felt great to be back on the road, even if this was a busy highway that led directly into the aftermath of a volcanic eruption.

CHAPTER 24 - A HIGH POINT

DISTANCE TO BUENOS AIRES: 6,133KM

It felt strange heading south knowing that I'd be returning to Quito by bus in just a few days but if I was to reach Buenos Aires by News Year's Eve then I had no choice. With 6,000km and just 135 days left, my daily average had crept up to over a marathon and logistical issues were stacking up against me.

Putting the pressure of daily averages aside, I was desperate to get as close to the recently erupted Cotopaxi as possible while the intensity of the eruption could still be experienced. The further south I pushed, the more I could see the layer of ash that had formed everywhere.

It is embarrassing to admit that I spent most of the first day focusing on what I thought was Cotopaxi, only to find out it was a completely different summit. It wasn't until I spoke with the other guests at a climbing hostel in Machachi that I realised my mistake. Thankfully these tourists, both Ecuadorian and from overseas, were perfectly clued up on events in the surrounding peaks.

The running during this section of my journey was exhausting. Not only were the many trucks and buses spewing black smoke as the road weaved its way over long, high climbs, but I was running at altitudes of around 3,000m and that had a real impact on my breathing and performance. Every kilometre became a struggle and I found it increasingly hard to appreciate my surroundings. Out of the corner of my eye, the horizon was punctuated with mountain peaks soaring on either side, but any loss of concentration left me open to all the dangers the road presented. Keeping my head down, I just had to get the kilometres done, ensuring that my arrival in each village or town was sufficiently early to find a cheap hostel to rest for the night. The further south I forged, the more built-up the road became, as half-finished buildings clung to the highway and small businesses sought custom from passing traffic. All of this, along with the knowledge that the further I ran the longer the journey back to Quito, drained my motivation. I tried to put these issues out of my mind but couldn't shake the feeling that until I was free, I was essentially trapped.

The stretch of road from Ambato to Riobamba will stay with me for a long time for it was here that I first set eyes on Chimborazo. This mountain is famous not only because it is the highest in Ecuador, but its summit is the furthest you can get from the centre of the Earth while remaining on land, and therefore the closest place to space on Earth. For the entire day, this huge volcano dominated my view and somehow mesmerised me. Its imposing majesty distracted me from all the other things in my life and just held my attention. At that time, I couldn't have known that in 2020, just before the outbreak of COVID-19, I would return here and successfully make it to the summit.

As I arrived in Riobamba, the all-clear came to head back to Quito and retrieve my package. On securing a place for my stroller, I found a bus back to Quito. Despite having spent so much time in Quito already, it was surprisingly good to be back there, and relief washed over me on re-entering the post office and being rewarded with a box containing everything I would need for the next stage of my run. Back at my hostel room, tearing the box open had all the excitement of Christmas! I was delighted to find that my mother had added a few extra items, including saucisson, sweets, and gels but most importantly my new shoes. My two existing pairs had all but worn through and I could feel my legs jarring at every step. That night I enjoyed one last calzone and a night of socialising with a group of tourists. I hadn't been drinking for over forty days and maybe being more focused allowed me to enjoy my time with other people, rather than the inebriated interactions I had in North and

Central America. Whatever the reason, spending time with these travellers felt particularly restorative and unknown to me at this time, these new friends would resurface at numerous other pivotal points of my adventure.

I don't know if it was the new shoes or no longer having to wait around, but my new lease of life was unmistakable as I left the Spanish colonial architecture of Riobamba and started my push towards the Peruvian border. With the mayhem of the highway behind me, my route cut into the mountains, meandering from the built-up valley floor into a greener, more exciting landscape. This day would mark the highest point of the expedition so far, at 3,850m above sea level. Just beyond the pass, I found a small area off the road to camp, foolishly not considering how far the night-time temperatures would plummet when so high in the mountains.

There's no doubt that it was the extreme cold that woke me the next morning. Everything was frozen, which made everything that little bit more difficult, but rather than dampening my spirits, it whet my appetite for the challenges that the Andes in Chile were going to present. Any lingering worries soon evaporated when I started a descent that plunged nearly 2,500m over the course of the day and marked the transition from mountains back to coastal roads. This valley was a deep, intoxicating green and huge trees were scattered on its steep sides. Piercing this fertile paradise was the jarring yellow line down the centre of the road as it flowed downhill and pulled me with it. Thankfully, the gradient was just at the point at which I could run freely without jarring my knees to control my speed. The pay-off was a heady dose of runners' high, to which I had become hopelessly addicted.

My descent from the mountains continued through Pallatanga, Cumanda, La Troncal and Puerto Inca, where I re-joined the highway that would take me to the border. As banana plantations replaced mountain crags, the surroundings and temperatures changed so dramatically in such a short time, such that humid conditions made running sweaty and a little uncomfortable.

Under some illusion that crossing the border into Peru would change the running conditions, I put my head down for the next

150km to make up for lost time. The flatness of the scenery made it seem bland after mountain running, and only the odd low-flying crop-duster overhead broke the monotony of the long, straight roads.

Near Santa Rosa, banana plantations gave way to cocoa. As I passed one such plantation, one of the workers who was tending to the trees beckoned me over and offered to give me an impromptu tour. As a group of workers watched my stroller, I was shown the whole process from planting to harvesting. My guide plucked a pod from the tree and broke it open to reveal the succulent white flesh-covered beans and he invited me to suck one. The sweet taste gave me the sugary hit I needed, instantly reviving my mood. Full of energy, on emerging from the trees I failed to notice the low wire fence around the trees and, mid-conversation, went tumbling right in front of the audience that had grown around my stroller. Needless to say, this provided a huge laugh to all, me included.

Arriving at a border always presents issues but never more so than when you run to the completely wrong place. That is what happened at the estuary town of Huaquillas. After a longer-than-marathon day, I blindly headed directly to the town centre, past the Monumento a la Paz (Peace Monument) and arrived at a vibrant and colourful part of town where I expected the border to be. However, some years previously a new border crossing had been established around 8km away and back the way I had just run. Evening was fast approaching, and I was nowhere near prepared for the crossing, so opted to spend one last evening soaking up the Ecuadorian vibe. Sitting on colourful plastic furniture next to brightly lit food stalls, excitement welled up about Peru, the country I would spend the next 69 days crossing.

The next morning, back on the Pan-American, I reached a new complex designed to process greater numbers of people and trucks. Long queues snaked around the courtyards, making it impossible to wheel my stroller into the offices as I had at previous borders. I found a friendly-looking policeman and asked where to leave it and he pointed to an open space between the Ecuadorian exit and the Peruvian offices. Feeling uneasy about leaving everything I owned in the open, I tried my best to keep an eye on it throughout the admin process.

It took an age to be processed into Peru, so I decided to limit my aim to Tumbes, the first main town on my way south. Here I could re-stock, get some local currency and find my twelfth SIM card of the expedition.

CHAPTER 25 - ESCORTED BY THE POLICE

DISTANCE TO BUENOS AIRES: 5,502KM

Tumbes is a busy port, but also attracts a lot of tourists so finding a hotel wasn't difficult. The town itself has charm and has a beachy, seaside resort feel. Shopping arcades line the streets, from which vendors sell all sorts of souvenirs and items, presumably to appeal to Ecuadorians on brief sojourns over the border.

With a strange rush of elation, I felt the presence of the Pacific Ocean on my right as I took the road south out of Tumbes. The ocean had been my constant from Canada to Panama City and had kept me true throughout my journey. The mountains of Colombia and Ecuador had replaced it in recent weeks, and I drew great comfort from returning to the sea once more.

Little tourist villages and quaint hostels peppered this coast, tempting me to stop for a few days. But with 5,500km still to cover and only 125 days remaining to arrive in Buenos Aires on time, frivolous stops were hard to justify. I am only human, so allowed myself to stop for a delicious seafood lunch in Zorritos, although the price was some serious mileage that afternoon. After 65km, I arrived at a small hotel run by a Peruvian-American couple who took me in and fed me up for the next day.

Mancora, a well-known destination on the backpacking trail, was next on the route south, and having run nearly 530km over the last 11 days, was a useful resupply point. Curiously, and perhaps out of some subconscious desire to be among other people, I booked into the Loki hostel, renowned for its partying and less so for catching up on sleep and recovering. This would be a day off, so I set about having the most relaxing time possible.

Pulling together my washing, I made my way to the laundry in town. The first local who saw me exit the hostel approached and asked if I wanted to buy cocaine. This was repeated a few times on the short walk to the store by various guys on bikes and even inside the store. I was quickly getting a clearer picture of what this place might be about.

Even amid this hedonism, I kept myself to myself. As a non-drinking adventure runner, this was not exactly my tribe, so I used this time to do admin and eat as much food as possible instead. However, sleeping proved patchy with the raging party outside and once again a roommate's continual need to access his cocaine stash in the drawer next to my bed!

I made friends with one of the guys who worked on reception and together with another couple of Peruvians, we set out for breakfast. It was nice to get away from the touristy part of town but even so, I was back in that weird cycle I seem to enter where I crave company and then immediately reject it.

Another reason for stopping in Mancora was the upcoming change in the environment. The Atacama Desert was prominent in my consciousness, but I was unaware that Peru had its own deserts. The main one in the north is called the Sechura Desert and it starts near the town of Piura, a few days south of Mancora. I had become quite good at picking up local intelligence as I ran, but that the next few months would almost exclusively be spent running in desert conditions had so far eluded me. My only real training for this harsh climate was the time I'd spent in the Baja region of Mexico. New challenges included dealing with extreme heat, a lack of shelter and of water, the limited civilisation of any kind and a heightened chance of encountering banditos. My time in Mancora was therefore vital to mentally prepare for what lay ahead. Ever since arriving in Latin America, I had been warned about bandits, but here that threat seemed more real, and even more so on receipt of this email from a member of the Pedal South team I had met in Mexico:

‘I was robbed by three armed gunmen recently on the highway.... This was in Northwest Peru just south of a town called Piura. Apparently, many cyclists have been robbed on the same highway. It was scary, a lot of gunfire. I had to fight my way out and jump onto a big truck that slowed down for me. They did not hesitate to shoot the heck out of those guns, both to scare me at first, and then shooting at me as I escaped on the highway.’

As warnings go, this seemed bad. Piura was only 200km south and my fear levels were rising for the first time. All I could do was believe in myself, act responsibly and keep vigilant. Above all, I couldn’t let these warnings dictate the journey. I had been running for over a year and had to have confidence in my growing skills and ability to adapt.

Getting back on the road was a little tricky, as it always is after a day off, and to make it harder the road south of Mancora was quite undulating. Plus, the dramatic change in the landscape took some getting used to, as the greens had by now disappeared leaving only desolate, harsh land. The distances between little settlements grew and resupplying became more complicated. On one occasion, a large lorry pulled over and its friendly-looking driver peered down at me from his cabin as he leant across to his passenger’s seat, grabbed a plastic bag containing his lunch and lowered it down to me, before smiling and driving off.

Stealth camping also became harder and most of the time, near impossible. The desert was pancake flat in all directions and I had to become more creative when setting up camp in the evening. In this exposed environment one tactic worked, namely to wait till the last possible moments of daylight to pitch camp and then keep any lights off as soon as it got dark.

The next morning something didn’t feel quite right and unfortunately, the trucker’s meal might have been the cause. On reflection, a hot meal sweating in a plastic box for a whole day is probably a breeding ground for bacteria. There are a few issues when running in the desert with a bad stomach. First, diarrhoea dehydrates you and flushes out the nutrition you need to continue. This, combined with the heat of the sun and lack of shade, results in chronic dehydration. Finally, there is the remoteness. No matter how I felt or what I wanted to do, the only option was to continue moving no matter how unpleasant that was. Sitting around in a desert is not an option, especially with limited water supplies.

One benefit was that the dramatic scenery helped make the struggle more bearable, as a huge canyon cut the desert landscape in two. The Pan-American took me through the extraordinary scenery before diving down to the canyon floor below, where the air was stifling. Small huts on the side of the road sold basic supplies, including warm cans of Coca-Cola that boosted my energy levels while potentially killing some of the bacteria in my stomach. The medicinal qualities of Coca-Cola may be far from proven but they definitely had a placebo effect for me.

I struggled on, choosing to find somewhere to sleep near the highway rather than divert to the coastal town of Talara. It was clear that a couple of slightly run-down huts on the side of the road were being inhabited by a couple of families. I asked if I could camp nearby and they showed me to a flat area where I could pitch my tent. The next thing to secure was food supply; I asked at one hut if I could buy any from them. The hut had just one room, where both parents and children slept, cooked and ate. The man pulled back a rug to reveal their store of food, which mainly comprised sugary drinks, biscuits and sweets. I bought a selection, thanked them profusely and headed back to my tent.

The next day, my strength had returned and I managed to cover the 55km to Sullana, which sits on the banks of the Chira River. The water from the river irrigates this region and farmers were growing crops on every inch of available ground to take advantage of the fertile conditions. Paddy fields dominated the land, beautifully reflecting the images of nearby large trees in the still water. What a contrast to the surroundings I had become accustomed to over the last few days! The abundance of water also attracted people, making the town itself a vibrant and bustling partner to the serene river below.

In a nice change of routine, I had made good time that day so the afternoon was free for a little exploration. I hailed a tuk-tuk into town and visited the museum for an update on the rich history and culture of the region. Previous civilisations of this area include the Moche (100-700AD), the Chimor (900-1470AD), the Incas (1470-1533AD) and the Spanish. The town, as it is today, was founded in the 18th century and the architecture, while run-down, was worth exploring. But for me, one of the biggest attractions was the Tottus supermarket that dwarfed nearly every supermarket I had ever seen. It seemed a mecca for the locals, most of whom, like me, were there to enjoy the air conditioning.

Piura was the next town on the map and as I ran towards it the words of warning from Pedal South echoed in my mind. To its south lay a 200km stretch of road that I categorically knew was dangerous and however much I tried not to let it affect me, clearly it was.

Although not a huge believer in higher beings, for a moment that day, I could have been persuaded. While mulling the dangers that lay ahead, I recognised a voice coming from behind me. When I turned around, I was both stunned and overjoyed to see it was Brooks, the American cyclist I had met in the US and in Mexico. This was our third meeting in just over a year and this time we were about to embark on a particularly perilous stretch of road. We chatted briefly before Brooks left to sort out some accommodation in town.

Over dinner we discussed the road ahead and assessed the dangers it posed. Brooks wasn't 100 percent sure he was leaving the next day, but I was determined to push on, so we agreed that I would set off and he could catch me up. At that, we polished off our meal and then headed into town to join the locals enjoying the balmy night around the town square.

The next morning, after a hearty breakfast I set off south towards the infamous Sechura Desert. As I made my way through the backstreets of Sullana, I was conscious of eyes upon me. Perhaps it was more noticeable than normal purely because I was more aware, but it definitely made me feel nervous. The feeling intensified when I stopped at a small store to stock up on some last-minute supplies and everything in the shop was behind iron bars. To buy anything you had to instruct the teller what you wanted and then pay before the goods were handed to you. My uneasiness only increased as people started to ask what I was about to do and where I was going. Every mention of running to Chiclayo was met with disbelief and sometimes very animated responses that ranged from, 'you must be mad,' to 'you'll get killed.' Undeterred, I pressed on, trying not to let these warnings get to me and left the confines of the city onto roads that seemed just like the road I had been following for thousands of kilometres. My nerves calmed, but only for a couple of kilometres.

I hit a police checkpoint and a small cluster of restaurants and shops by the road. Once again, everyone's eyes were on me. What did this crazy gringo think he was doing, their looks seemed to say. Young girls shouted at me while crudely running their fingers across their throats. I didn't have to be a genius to interpret what they were hinting at, so I approached one of the officers to ask if it really was as dangerous as everyone was making out. They seemed unfazed and, in an unconvincing way, tried to pacify me by ushering me on my way.

At this point, I just wanted to get running. Every minute wasted was only adding to the sense of danger and there was no escaping this stretch of road. Unbelievably, this was the first time on this expedition that I had started to worry or question what I was doing. Although happy to take the risk, was this fair to those who cared for me back home such as my parents, siblings and girlfriend? Was it fair that I was potentially risking everything and if I was risking everything, what was I risking it for? Thoughts like these can turn into doubt and considering my current predicament, they had to be pushed firmly from my mind in favour of what needed to be accomplished. I was going and there was nothing to be said or done differently.

Nothing really happened for the first few kilometres and the more I ran the more I put all the worries aside. After 11 different countries there was no reason why this small stretch should be any different. The small settlements that lined the road slowly disappeared and the desert took over.

Just as I settled into a rhythm, a police car pulled alongside and flagged me down. The two officers I had been talking to at the checkpoint jumped out and approached me. At first, I worried they were going to tell me not to run, however, it transpired that on reflection they'd thought they'd better escort me across the desert to a camping spot and then return the next morning to continue. This was clearly not a negotiation, and I secretly felt relieved. Just as we were preparing to hit the road, miraculously Brooks rolled up on his bike and, on hearing the arrangement, decided that he was going to share the escort. Together, we would make sure we got through this section of desert. While I enjoyed tackling the more challenging stages of my adventure alone, my connection with Brooks was such that I actually enjoyed having his input and positivity with me on the various occasions fate had us meet.

Our new escort set off and then would pull over every few hundred metres. Once we had passed them they would leapfrog us, which continued throughout the day, though it became apparent that they didn't want to hang around and they set a mean pace. After more than 60km we arrived at a rundown restaurant by the roadside that was used as a stopover for lorry drivers. The police indicated a patch of concrete and instructed us to sleep there and nowhere else. As they left, they shouted that they would be back at 7am, at which point we should be ready to leave.

The night's sleep was far from ideal, with large lorries arriving at all hours, their bright headlights blinding us as we lay in the open on inflated mattresses. With no shelter from the sunlight, we were up early, making sure we had eaten breakfast and were ready before our escort arrived.

The rapid daytime pace continued from the get-go and while the desert heat was punishing, we were soon making good time. At around lunchtime, we spotted a restaurant and offered to buy our escorts some lunch, which also ensured that I got a bit of a rest before we continued.

It was after lunch that our protectors informed us that we were at the border of their jurisdiction, and we would be met by another escort soon. It did feel a little like we were being abandoned, although we had not experienced anything remotely threatened, so felt confident to continue. A little after lunch a cop car did indeed turn up, but this one took a rather laissez-faire approach to escorting. For these officers, it sufficed to tell us they knew about us before disappearing.

In spite of being preoccupied with the threat of imminent danger, running in the Sechura Desert was truly breathtaking at the same time. Sand encroached onto the tarmac of the arrow-straight roads as the desert stretched out in all directions. Yellow road signs announced this was a 'ruta de dunas', which would have been evident from the wind-sculpted trees in any case. Running in such an exposed place was exhilarating but it did pose a big problem—where to camp?

While discussing our plan, another problem arose. Brooks's derailleur had caught the spokes of his rear wheel and had caused a huge amount of damage. Although he was unhurt, we were definitely more vulnerable now. Neither of us could work out how to make the bike rideable and to make matters worse it was getting late, so whatever we were going to do had to be done fast. A motorcyclist pulled over and tried to lend a hand but with no success. The last resort was to flag down a van and get a lift. Brooks urged me to continue but I didn't want to leave him stranded. Thankfully, a passing van pulled over and kindly agreed to take Brooks to a restaurant a little further down the road, where we agreed to meet.

I was now alone in a desert that was reputedly dangerous and the light was fading fast. To add to this, I was getting tired after running over 250km over the last five days and my legs were starting to feel less steady. Dim lights on the horizon gave me hope but when I got there, there was no sign of Brooks and to say it looked a little shady would be an understatement. I struggled on, feeling the power leaving my legs until finally, I saw the outline of a far more inviting restaurant.

I stumbled through the door to find Brooks sitting at the table sipping an ice-cold drink. Clearly, there was no reason for him to be here, other than to keep me company as the driver had been going all the way to Chiclayo, the next major city and the only place

Brooks would have found spares for his bike. Spotting my tinge of guilt, Brooks assured me that he wanted to make sure I was safe. These were some of Brooks' great attributes—dependable and always looking out for others. A true gent.

It wasn't long before food arrived on the table. I remember a momentary suspicious glance at the meat but after 78km you will eat anything that is put in front of you. The owner showed us where we could pitch our tents and soon it was lights out.

As soon as I woke up the next morning, I knew something was wrong. Beyond lacking energy, something else was nagging away and after I made the trek into the desert my worst fears were confirmed—diarrhoea. Taking a day off really wasn't an option because it was this restaurant's food that was the main suspect for my stomach issues. But it meant pushing a baby stroller more than a marathon distance to the next settlement through a desert, and to do all that with food poisoning.

Brooks and I agreed to meet up again in Chiclayo. However, after only a few kilometres I had to run into the desert once again and every time I attempted to eat or drink anything my body would react violently. Despite everything, my body still needed energy and hydration, so I had to balance what I took on with how often I was prepared to run for cover. As the day progressed, the distance I was dashing from the road grew increasingly short and before long I was just dropping my shorts at the roadside and casually waving at passing buses. This day was horrible in so many ways, but I doggedly pushed on until the desert was slowly replaced with land that had been reclaimed for agricultural use. This meant a settlement was getting close.

Mórrope is nothing grand, with dusty, deserted streets, dilapidated buildings and shops with empty shelves, but it did have a hostel and that is all I sought. I hauled my stroller up a long flight of stairs and was shown to a room where I collapsed, feeling sorry for myself. I showered, rested and did some minor maintenance on my stroller before heading into town. While I couldn't keep any food down, it was still imperative to keep the calories coming in.

The next morning, I had two choices. Remain in Mórrope and hope to get stronger or struggle the last 25km to Chiclayo. My unwavering desire to keep moving forward made the decision for me and I dragged myself out of bed and back onto the road. Shattered and drained on arrival, I managed to make contact with Brooks and soon we were reunited. Chiclayo is a small city and after a gruelling 400km over the last eight days, it seemed fitting to take a couple of rest days to recuperate. While Brooks may have managed to get his bike fixed, he had not avoided the atrocious stomach bug and unfortunately for him, was bedridden.

CHAPTER 26 - THE CHAN CHAN IN PERU

DISTANCE TO BUENOS AIRES: 4,961KM

I roamed the streets of Chiclayo, the fourth largest City in Peru, alone as Brooks recovered in bed, as well as making the necessary repairs and restocking on essentials. After just one day I was already feeling restless and it took all my resolve to take an additional day's rest and regain my full strength after running two marathons through a desert with food poisoning. Despite eleven unplanned days off, if I pushed, I could still be on target to reach Buenos Aires by New Year's Eve. I had 111 days to cover just under 5,000km, but still had the biggest obstacles to come—the Atacama Desert and the Andes. With the challenge still very much alive, although getting tougher, I couldn't be held any longer and with Brooks's assurances that he was well enough, I made my preparations to continue.

The desert directly south of Chiclayo was epic. On the right-hand side of the road, the sand stretched out to the Pacific Ocean and on the left, it ran to the foothills of the Andes. My immediate problem was what lay between, however. A relentless wind coming off the ocean created a carpet of sand that flowed down the asphalt. The wind would propel my stroller sideways and the sand would fly up into my face, making every kilometre a struggle.

A passing motorbike turned out to be conveying a young couple with some useful information. They told me about a restaurant called La Balsa, which was run by a chap called Clemente, also known as 'The Angel of the Desert'. While it was still some 440km away, this gave me something positive to focus on.

As I was still trying to regain full strength, I decided to cut the day a little short and stop in the small village of Mocupe about 40km south of Chiclayo. It was very basic but had everything I needed to ensure decent rest for the day ahead.

Feeling stronger the next morning, I was up early and determined to cover as much ground as possible. The beach town of Pacasmayo was about 60km away and seemed a likely target. With perfect running conditions over a nice flat road and under an overcast sky, I was feeling so strong that every time I reached my intended rest spot, I would push on to take advantage of the rhythm. Forcing myself to stop in the charming village of Guadelupe, there was a lovely restaurant on the central square where the equally lovely owner plied me with free chocolate ice cream. A little reconnaissance into the route ahead uncovered a shortcut from the Pan-American that would cut out Pacasmayo and arrive instead in the small town of San Pedro.

It was refreshing to be running on smaller roads and away from the monstrous lorries. The scenery was also less tainted by human development. I happened across a family trying to push start their truck so lent a hand. Although they were bewildered, they also seemed very grateful.

It was late evening when I arrived in the town of San Pedro needing somewhere to rest. Despite the town being quite large, the options for accommodation were limited to a place by the name of Hospedaje Delicias, which I figured would be more of a by-the-hour type of establishment. I scoured the streets and asked around but nothing seemed available or within my budget. My only choice was to chance it with Delicias. The ugly, half-painted pink building was on the outskirts of town next to the main road and behind large metal gates. It looked as seedy as it sounds and nothing about the place filled me with confidence. The girl behind the counter seemed astonished by my appearance and even more so when I asked how much it was to rent a room for the whole night, for just one person. At this, a young man appeared from the office and quickly took control. The price wasn't unreasonable and I took a room on the ground floor, which came with a discreet garage where customers' cars would be hidden from passers-by, other guests and probably most importantly, their partners! Inside was an enormous bed enclosed by bright pink walls and a disturbingly large ceiling mirror.

As I prepared instant noodles and tuna on my small stove, I tried very hard to barricade the thoughts from my mind of what may have happened in this room before. Fortunately, having run 67km, it wasn't long before I was fast asleep.

A further 100km south was Trujillo, which I was confident I could make in just a couple of days. This wouldn't be my first visit to this city and experience told me to take a day off to explore the Chan Chan ruins. For a more relaxed vibe and instead of staying in the main town, I opted for the nearby surf town of Huanchaco. The run there was mostly through desert but as I got closer, it gradually changed to a greyer, rocky landscape. Large mountains rose out of the sand and small, seemingly deserted rectangular buildings began to appear on the roadside. These tumbledown shacks were soon accompanied by large piles of rubbish strewn along the verges on the outskirts of the town. People rifled through the larger piles to see what they could find of any value, while stray dogs foraged for anything that was edible. These poor creatures looked at me with pleading canine eyes as I ran past. It was hard to translate whether the look was, 'please help me' or, 'please end this for me'. Either way, it tore me apart each time.

Luckily, Huanchaco itself was far more pleasant and once again, the restive presence of the sea so close was reassuring after the harrowing sights on my way there. Huanchaco is a hotspot for backpackers, and among the hostels I found a small place called Casa Fresh, run by an Irish chap called Richy.

The next day I split between visiting the Chan Chan ruins and enjoying Huanchaco. The village alone was very interesting and its coastal position means it has a rich and varied history. During the Moche, Chimor and Inca periods it was a major port and ever since has been a fishing town with local fishermen still using the long reed boats known as caballitos de totora (reed horses) that have been in use for thousands of years. Some believe it to be where ceviche was invented and that was the only excuse I needed to eat as much as I could.

The UNESCO World Heritage site of Chan Chan, meaning Sun Sun due to the sunny conditions, was the largest city in the pre-Colombian era of South America and the capital of the Chimor Empire from about 900 AD through to around 1470 AD, when it became engulfed by the Inca Empire. The site itself is hugely impressive and the visible structure is only a fraction of what could still be hidden beneath the sand. The walls are constructed from adobe bricks, rendered smooth and lavishly decorated with carvings of birds, fish and small mammals. So many aspects of the ruins captivated me, such as the advanced irrigation systems, and the way that the walls are designed seemingly to cater for potential seismic activity and protection from the heat, wind and potential El Niño.

Back at the backpackers, I got to know Richy a bit better and loved learning about how he ended up so far from Ireland. He had been an accountant before realising that there was more to life than sitting behind a desk, so made the trek to Peru and was now running a hostel. His attitude and determination were awesome and he had one of those infectious characters that meant everyone around him loved him.

Determined not to get stuck, I was back on the road the next morning and for a change, this time had company. Richy and his friend Kevin decided to join me for the first 15km of the day, which took us to the other side of Trujillo. Both took it in turn to push the stroller and, I think I can safely say, were slightly surprised at the pace I was trying to maintain. We stopped at a small café where they bought me breakfast before jumping in a taxi, leaving me to run another 50km.

Refreshed after a day off and good company having lifted my soul, it was no wonder I felt a strong determination to make this a big day. As I'd made light work of my original target and was feeling so surprisingly fresh, I reached for my map and decided to make the extra push to a village called Virú. But on arrival, the name on the map was clearly not the actual town but more a junction with a collection of shops and restaurants. The mood was off and I did not feel comfortable camping there. Returning to the map, I found that the real town was about 4km away from the main road and that meant digging deep to find the extra motivation and strength to make it there. I just kept telling myself that there was a bed waiting for me after cramming 65km into one day.

Virú was a pretty little village and I'm glad I made the detour. My hostel was just off the main square, where a market took place during the day. Although its wooden stalls with makeshift roofs, dirty streets, assorted bright tuk-tuks and beaten-up cars were the wrong side of untidy, candles were lit in the stores as the night drew in and the darkness and shadows transformed it into a beautiful scene. As with many South American towns Virú's central square was in constant use with children taking part in dance lessons and sports. Stopovers in a town like this make the long runs and detours worthwhile.

The next morning, I found an alternative route back to the Pan-American and resumed my way south. This road cut through an agricultural area and old lorries came storming past with seemingly impossible loads of sugarcane strapped on. By midday, the sun had reached its zenith and the heat was becoming unbearable. In the small market town of Choa, I found a small restaurant where I could eat and shelter from the sun. Possibly because of the harsh conditions, it took me longer than normal to regroup and set myself the target of reaching a town called Guadalupito, which didn't look too far south. Re-enter the desert I did, and mustered all my strength to maintain my running despite the increasingly difficult environment. The road wound up a huge hill and through a man-made cleft in the rock at the top, where large dumper trucks and bulldozers were forging a new motorway through the mountains. The problem for me was that this added complexity to the running. Fortunately, everything seemed to change at the summit, and as I let gravity pull me downhill I could feel my body recharging and my mood assuming a more positive aspect. The run that followed sticks in my mind as one of my favourites of the whole expedition. The valley carved its way down towards a vast plain below where the flat valley floor provided the most spectacular natural arena. Running on Empty from Forrest Gump played on my speaker at full volume as I joyously sang at the top of my voice. Some days start tough, then quickly transform into euphoria.

To get my bearings, I pulled into a small mixed-use dwelling and shop. The sun was starting to set and finding anywhere to shelter from the desert winds was proving difficult. Buoyed with positive momentum, it seemed within reach to push on a little further to make it to Guadalupito, which the shopkeeper assured me was less than 20km away. This turned out to be another instance of local people the world over having no comprehension of distances as they base everything on minutes in a car.

It would be some time before I discovered the inaccuracy of the advice I had received, however. After a large hill the road opened out onto a huge plain that presented no opportunity for camping and gave no hint at any civilisation as far as the eye could see, making my decision to continue a little enthusiastic perhaps. But in these moments of potential disaster, a new determination kicks in and so I just put my head down, picked up the pace and raced the sun that was setting over the Pacific. Having run over 60km for the second day, I spotted some lights in the distance and fixed all my efforts on reaching them as my legs started to burn. The lights shone from a sort of tourist recreation park with slides, swings, pools, and brightly coloured play areas, and they seemed to swirl and flash in my current semi-delirious state. Of course, the park was closed but I managed to find someone working there and asked about staying the night. He assumed I meant renting a cottage and quoted a ridiculous price. It was becoming evident that neither of us understood anything the other said, probably down to a combination of local dialect and my still embarrassingly atrocious Spanish. Luckily, the arrival of another guy untangled the miscommunication, and I was invited to set up camp at a crazy looking play area full of colourful plastic rides and fake grass.

The next morning, a sense of decency compelled me to buy my breakfast from the restaurant. If I was expecting pancakes or eggs, then I was in for a surprise. The menu was very much directed at families on a day out and mostly consisted of fried meats and chips. Thinking of my stomach, the blandest option seemed prudent, which was the equivalent of a very large fried chicken meal, complete with a Coca-Cola. The deep-fried fly in my meal was a little unsettling but was swiftly trumped by the huge rat that scampered into the kitchen, chased by one of the chefs trying to trap it. It was time to leave.

To arrive at Clemente's restaurant, La Balsa, at the end of a running day I had to work out where the best places to stop in between were. As I arrived on the outskirts of Neuvo Chimbote a young Peruvian couple on a motorbike drew up alongside. 'Where are you heading tonight?' they enquired. They presented no risk here, so I told them I had no exact plans for that evening! 'Well, ok, follow us!'

They weren't to know that I had already decided to stop early but at times like this, a lot of things run through your mind. First, there is the selfish thought of whether you want to spend the day with strangers and a language barrier when you are physically and mentally drained. This may sound ungrateful or even unadventurous, but to keep going day after day involves a lot of introversion and protective mental armour, a distraction like this can either be good or disastrous. Second is the need to judge instantaneously whether the people you are going to stay with are indeed trustworthy. In this instance, I had a good feeling about Alexander and Melissa and told myself to stop being so guarded. What was the adventure for, if not living in the moment?

We formed a convoy of two and crawled alongside the highway, passing through the town of Chimbote before arriving in Neuvo Chimbote. Alexander directed me through the streets to a small courtyard where he knocked on a door. When a lady called Ivone answered the door, it dawned on me that Alexander never intended to host me at his house but that of his friend's mother. Suddenly this felt incredibly intrusive, but my presence didn't seem to faze Ivone in any way. I later learnt that her house was part of an informal motorbike network that ran throughout Peru, designed to give motorcyclists places to stay along their journeys. Ivone did a little tidying and showed me to my room before Alexander and Melissa whisked me off for lunch at a local, very popular, ceviche restaurant. We shared stories of travel and discussed the differences between life in the UK and Peru, simply enjoying each other's company. After lunch, Alexander, Jorge (Ivone's son) and Giuli (Jorge's girlfriend) and I sped off on motorbikes to the supermarket after I mentioned a need to resupply.

Back at the house, I was left to my own devices until dinner when Ivone generously cooked up some rice and eggs. The next morning, after a huge breakfast and an extended selfie session with everyone who had been so hospitable, it was time to hit the road again.

CHAPTER 27 - ANGELS OF PERU

DISTANCE TO BUENOS AIRES: 4,638KM

Bolstered by the positivity and generosity of Neuvo Chimbote, the desert seemed somehow less harsh the next day. On the outskirts of the next village a group of kids were trying to fix a puncture, so my recent plentiful experience at tyre repairs came in handy. The desert opened up, reminding me of scenes from Laurence of Arabia and Star Wars so much so I put both films' themes on loop to complement the atmosphere. After 55km, I arrived at the town of Casma, also known as Tierra del Eterno Sol (Land of the Eternal Sun), with Bon Jovi blasting from my speaker. As with many of the towns in this arid region, Casma was built next to a river and offers a little reprieve from the desert, with canyons cleaving the rock and green vegetation sprouting from the soil. However, the desert quickly takes back control and on leaving Casma the black asphalt sliced through large sand dunes on each side of the road, providing a dramatic backdrop to run through. At a small roadside store, an old couple sold sun-baked glass bottles of Inca Cola, which I used to wash down some dry biscuits.

It was 30km before I spotted an orange building with 'La Balsa' written on the side, which translates as 'the raft'. This was the home of the local legend Clemente, or the 'Angel of the Desert'. Its windows were struggling to perform their main task of letting light in, so covered were they with stickers from motor bikers, cyclists and adventurers who had ventured into this haven. On enquiring if Clemente was at home, a pretty lady, who turned out to be Clemente's daughter, introduced me to a smartly dressed gentleman in a white shirt and grey trousers. Just by looking, you could tell he was a kind person. He offered me a seat and asked what I would like to eat. Soon I was tucking into a huge plate of lomo saltado, a succulent Peruvian beef dish in a beautiful sauce accompanied by chips, rice and an egg. The hospitality here was so warm and welcoming that when I asked if I could set up a tent, Clemente just smiled and showed me to a small room out the back with beds, a toilet and shower. He had specially created it to give weary travellers a place to sleep for the night. After a wash and change of clothing, I returned to the restaurant to learn a bit more about Clemente.

Clemente had been in the Peruvian navy and grew up around Lima. About 48 years previously he had moved up to the north and established the restaurant that is now called La Balsa. Over the years he had taken to offering shelter, with a special focus on motorbikes and cyclists. He had welcomed a few people on foot, but the numbers remained in the single digits and runners were something of a novelty for him.

Pouring me a coffee, he began to pull out the various notebooks, clippings and photographs he had compiled over the years. The language barrier was evident between us, but this didn't prove any impediment as the conversation was very one-sided, with me nodding along to his stories of warships and travellers. A big blue notebook turned out to be his visitors' books, through which he eagerly flicked, pointing out guests of note. I spotted a message from Scottish cyclist Mark Beaumont and knew I was in good company. The conversation embraced various topics, including who had first inhabited Easter Island, Polynesians or Peruvians, something Clemente was very passionate about. Nourished both by the chatter and a beautiful dulce de leche covered pancake, I went to bed with a full stomach and a happy soul.

The next morning, Clemente's beneficence continued with another delicious meal and a packed lunch to take with me on the road. Customary photos taken, we said our farewells. While the temptation to remain was strong, the desert was calling.

On the empty highway, the heat of the desert was oppressive and trying to find any shade proved near impossible. No matter what time it was, my stroller provided no shade from the sun, or the sand the desert winds whipped up into my face as I ran. Worse, trying to achieve continuity was proving hard, making running 50km a day difficult.

After 52km I arrived at the town of Huarmey with a very simple mission—to find a hotel with good enough Wi-Fi to allow me to watch the Scotland versus Japan World Cup rugby match that aired the next morning. Trying to find good internet in the desert is not always easy and, on this occasion, it demanded financial outlay. In all honesty, the spend was worth it as I was relieved to have a nice room, cut off from the world around me. The desert running had been exhilarating but the constant battle with the elements was wearing me down.

The next morning Scotland did me proud and beat a Japanese side who were fresh from taking down South Africa. Resisting the temptation of staying another night, I dragged myself away from the luxuriously air-conditioned hotel room and ventured out into the hot desert. For kilometres, the road was my only reference point, with nothing but the disorienting sand as far as the horizon in every direction. My only respite was a small village that straddled the road and provided a short relief of greenery until the restaurant 40km down the road I had targeted. When this otherwise nondescript and unremarkable establishment came into view amid the stark and barren landscape, the relief was overwhelming. Day by day, the desert was proving more and more demanding.

La Cabana was an apt name, as it was a small, circular building with a reed mat roof to provide shade. Venturing over with my stroller, I was hoping to find some food and a place to pitch the tent but was greeted with neither. The proprietors looked less than impressed to see a gringo at the door and very quickly asserted that they were closed and that camping was not available, leaving one option only: keep running. Having been so used to being welcomed in, it felt strange to be so coldly turned away.

After a few more kilometres I arrived at a motley group of shacks in which dilapidated buildings sat between restaurants so tired-looking that you couldn't determine if they were still operating. Nonetheless, hope was not lost as reed matting announced in black writing that there was 24-hour security here, which indicated there might be somewhere to pitch camp. With no other way of choosing, I entered the restaurant with the most lorries parked outside as an indication of its popularity and was rewarded by its surprisingly homely interior. Strong coffee flowed while lorry drivers watched TV on the corner screen. Mindful of the need to pay my way, I ordered coffee and dinner before asking if I could camp outside. A patch of ground behind the main building would have been ideal, but in fact my campsite lay through the kitchen, across the courtyard behind and in an empty potato shed with a wooden shelf attached to the back wall. My host moved a few boxes to make space for my stroller and then left me to set up camp, my mattress fitting perfectly on a wooden shelf. As rustic as it was and despite sharing it with rats as they scampered around, with exhaustion in my limbs the night's sleep was pretty sound.

The day's target was Barranca, where the plan was to meet a man called Cesar, who had contacted me through Facebook offering a place to stay. This was obviously really kind of him, but the desert running was getting to me by this point, so I half-heartedly agreed to just a drink that night, as all I really wanted was a hostel and to be alone.

The highway veered closer to the ocean and I could sense the proximity of civilisation. As this road was under construction, it was usual to see construction workers, but my eye was snagged by a group to my right wearing orange high-vis clothing. They were focusing intently on one section of ground so I pulled up and went to investigate. It turned out that roadworks had uncovered an ancient settlement and this group were there to excavate the site. The archaeological dig extended in all directions and was very evidently situated right where the road was planned to go. Catching the eye of one of the ladies, I asked what would happen when the excavation and chronicling was complete, and was dismayed to find that the bulldozers would arrive as planned and clear the path for the new highway. At first I was shocked that this was allowed and enquired why the road was not diverted to save the hidden town? The reply was simple. If they diverted the road for all the settlements that were found, then the road would zig-zag everywhere. It appears that sometimes the future prosperity of a country must take precedence over its past.

A short while later I came across a signpost to the Fortaleza de Paramonga. Having made good time and with only a few kilometres to go, a detour seemed manageable. At a small ticket office, I peered around the door to see a man asleep on the floor and not much else going on. He must have sensed my presence as he stirred and looked up at me, surprised that someone was visiting. He was a friend of the official guide and within a few minutes, I had paid my nominal entry fee and gained access. Apparently, the lack of other tourists was due to the fact that most people board the bus in either Trujillo or Lima and drive straight past.

Paramonga had been built in the 1200s and was on a hill at the edge of the Kingdom of Chimor. Through engineering ingenuity, the builders managed to channel water up from the local rivers to its series of circular walls and terraces. One chronicler, Pedro Cieza de León, passed Paramonga during his trip from the City of the Kings (Lima) to Trujillo in 1541. He described it as follows:

'There is one thing worth seeing in this valley, which is a fine well-built fortress, and it is certainly very curious to see how they raised the water in channels to irrigate higher levels. The buildings were very handsome, and many wild beasts and birds were painted on the walls, which are now all in ruins and undermined in many places by those who have searched for buried gold and silver. In these days the fortress only serves as a witness to that which has been.'

The guide explained that if you looked down on the fortress from a bird's eye view, you would see that is it is built in the shape of a llama. How they could have achieved this blows my mind. Today, graffiti covers the adobe bricks, some of which dates back to 1914. It is easy to imagine how impressive this building would have been at the peak of its power but now it seems little more than a forgotten ruin on the side of a busy highway.

Leaving the ruins, I made for Barranca feeling horrible but slightly relieved that I hadn't heard from Cesar. At a hostel near the centre of town, listening to the views of the incredibly religious owner was slightly disturbing, but a cheap bed is a cheap bed.

Cesar finally got in touch and insisted on coming to see me. Although apparently a little hurt at first that I had opted for a hostel rather than wait for him, that didn't stop him from offering hospitality in any way possible. Soon we were in his tuk-tuk and racing across town to his home and he was not taking no for an answer.

His house was built of a mixture of adobe and red brick, with green metal doors and reeds acting as curtains over the windows. To the right of the front door was a small adobe extension that housed a basic toilet. Cesar's house turned out also to double up as his workshop, as he was a mechanic by trade and all the space outside the house was crammed with spare parts and hunks of metal that to the untrained eye looked like nothing more than scrap. He ushered me into his house and introduced his kind-faced wife. Inside, the house was like nothing I'd yet witnessed on this trip. Space was very tight, with bunk beds on the right, one for the parents and the other for the two children. Clearly, privacy was at a premium under this arrangement. On the left was a small kitchen and table with the rest of the spare space used as Cesar's workshop. On taking all this in, I immediately wondered where it was that I would have been sleeping if I had stayed. A quick visit morphed into an invitation for dinner and soon I found myself sitting at the table with Cesar and family, sharing a simple meal of rice and meat. After an evening enriched by the warmth of this family's hospitality, if slightly devoid of fluid conversation, I thanked them and took a taxi back across town to my small hostel.

The next day would be a day off as I had run nearly 400km over eight days without a rest. My grand plans of catching up on admin were scotched when the local power grid had a different idea and with no internet, I spent the day strolling around the streets.

Before leaving Barranca, I returned to see Cesar with my stroller as he was keen to get a photo. But he wouldn't let me leave without giving me supplies and his wife searched the cupboards for things to give me. What really set South America apart for me was how kind and generous the people were and how insistent on giving whatever they have to help those around them. There are times when the success of my trip was made possible because of this quality. No matter how rude I felt, I couldn't bring myself to accept the evaporated milk and other gifts Cesar held out to me. I knew I would most likely not use them and their value was more to him than me. The only thing remaining was to come up with a story about why I couldn't accept his generosity and quickly extract myself from the awkward situation.

From there I next stopped at a curious little seaside village named Huacho. A peaceful night was necessary if I was going to make it from there to Chancay. This big push turned out to be 67km in the very isolated desert. Convincing yourself that you can indeed run that distance can be tough, especially when you ran 52km the day before and know another marathon awaits the following day. Trying to consume enough calories is also difficult. Before setting off, I prepared some porridge with jam, a cup of coffee and a cereal bar. With everything packed I was ready to go, before realising I was still starving.

To give a glimpse into my diet on the road, I will now list a day's rations in the desert. First, I visited a bakery and bought a couple of rolls, a pastry, and a biscuit. Once on the highway, at another bakery I added a couple of small croissants filled with dulce de leche to this horrendously bread-dominated breakfast. Later on, I feasted on cereal bars, biscuits and a couple of bananas. Lunch was a can of tuna with a piece of brioche. Then, at 58km I stopped for a packet of crisps, a Coke Zero and a small bar of chocolate for the energy to complete the last few kilometres. It's extremely hard to imagine all that in just a few hours now, but with the extreme exertion required, I had no choice.

Finally, Lima was in my sights. The prospect of staying with a friend of a friend there was very appealing, as well as the home comforts to enjoy. But with over 80km to go, breaking my journey into two days was inevitable. From where I was to Buenos Aires was over 4,200km and I had just 94 days until New Year's Eve. This meant that with no breaks, no days off and no injury days, I would have to run over 45km a day to make it on time, a number that seemed to be increasing daily. And while I desperately wanted to enjoy the luxuries of Lima, taking too much time off would be difficult to recover from.

My trip to the outskirts of Lima was a journey of two parts. In the morning, I headed south taking the quieter lorry road along the coast. What it lacked in traffic, it made up in hills and other unexpected obstacles. Along the way, locals stopped to give me oranges and the police pulled me over, though when they realised I wasn't a threat they left me alone. On local beaches families were enjoying time out, and I hurried past a shop that had twelve Rottweilers chained up outside. After passing a military base and the naval yard, I cut back inland and towards the bright lights and commotion of the second-largest city in South America. The impending city was making itself felt and a nagging tension was building inside me. Before tackling the pandemonium, I found a small restaurant on the side of the road to refuel and gather my thoughts.

As the city grew around me so did conflicting images. At one end of the spectrum was a huge new, sparkling, air-conditioned mall with every brand imaginable and at the other, a shantytown precariously fringing the steep side of a hill. It amazed me that despite, or probably because of, the evident poverty in this part of town, the inhabitants had taken the time to paint their homes in bright yellows, reds, greens and blues.

My chosen stop was in Los Olivos, where I may have looked somewhat out of the ordinary, running down a road surrounded by large billboards, restaurants, office buildings, parked lorries and general commotion. In the search for accommodation I found myself surrounded by American-style restaurants and cheap by-the-hour hotels. Determining which hotels would be appropriate was difficult and the more I plodded the streets, continuously battling my stroller, the less I cared about where I ended up. In the end I opted for a very clean and rather fragrant hotel that was quite clearly designed for activities other than sleeping. The indications were evident the moment you entered the hotel, with individual condoms and sex aids available for purchase at the front desk. But there was a bed and that was all I cared about.

The next morning I battled along the horrendous roads that bypass the airport. With no hard shoulder and far more vehicles than I was used to, running proved both difficult and dangerous. Finally managing to drop down onto the coastal road, at last the end of today's torment was drawing closer. As I ran, a small car pulled up next to me and out jumped a lady. It was Danica, my host in Lima, and although we'd never actually met, she greeted me like a long-lost friend. Danica was a friend of my good friend, Johnny, and the wife of his stepfather's brother, David. Not only had she generously agreed to put me up in Lima, but also to be my postal address for the latest care package. Her assurance that there was not that much further to go gave me the motivation I needed to cover the short distance to the Malecon de Chorrillos.

Arriving at a stranger's house in these conditions was sometimes difficult as my energy was low by the evenings, and while my hosts were excited, I was struggling to stand. But arriving at Danica's was incredible. She intuitively knew how I was feeling and showed me directly to my beautiful room, immediately giving me space to shower and get into clean clothes.

To say I had landed on my feet was an understatement. Danica's clifftop building had unrivalled views of the bay and the beach. The local private members' club was just a stone's throw away and Danica had managed to pull some strings with the powers that be and got me a guest pass. To add to that, Danica had organised a lunch with a couple, Wally and Patricia, with whom she worked as aircrew for American Airlines. I immediately liked them and unbeknownst to me, they would be amazingly helpful and generous once I got back on the road. That night I fell into bed and slept so well.

The next morning Danica was working, so had left me with free rein of her flat and the unwavering help of Maria, their maid. While our conversation was awkward, she kept a huge smile on her face and seemed determined to fill me with as much food as possible. The day before over lunch I'd mentioned the need for some spare parts for my stroller and to locate my care package from DHL. Wally had kindly volunteered to be my fixer, so I was thrilled when he arrived promptly after breakfast in his Volvo estate and we started out for the sprawling suburbs of Lima.

The first port of call was the post office, where it became apparent that getting my hands on my package was going to be as hit and miss as it had been in Quito. In a stroke of inventiveness Wally turned our excursion into a tour of the impressive Plaza Major after having to abandon his car briefly in a garage to cool down after a small overheating incident.

The rest of the afternoon was spent doing everyday tasks, such as picking kids up from school. Over lunch, Wally tried to introduce me to as many new fruits as possible, including aguaymanto, chirimoya and pitahaya.

Evening was designated to meeting up with Matt and Jess, whom I had met in Popayán, and we feasted on burritos in Miraflores. Seeing some familiar faces and hearing about other people's experiences on the road had a really grounding and joyful quality to it. I couldn't decide if I was slightly jealous of what they had done compared with what I was doing. Talking to travellers like these made me realise just how different my journey was. If ever I lost sight of my goals, these small interactions allowed me to put them into perspective and regain the drive to push on.

The next day was spent jumping into taxis to visit the various postal services of Lima. I'm happy to report that this time we successfully retrieved one package containing fresh running shoes and a whole host of products from France that my mother had added to remind me of home. As random as some of them were, they made me feel so happy it was to the point of homesickness. The second package, however, was proving far more elusive. Revisiting the DHL office the next day, it transpired that the problem related to customs restrictions and not having an address in Lima. Danica took it all in her stride and with her steadfast resolve, we finally obtained the parcel.

Just to prove that nothing is ever quite perfect, that night we returned to the flat to find that Danica's dog had taken a liking to the saucisson that my mother had sent me. Rather than waste the remaining sausage, we cut off the chewed areas and thinly sliced it for a dinner we were hosting that night. I promised Danica I wouldn't mention this—sorry!

CHAPTER 28 - ONWARDS FROM LIMA

DISTANCE TO BUENOS AIRES: 4,185KM

After four days in Lima and my packages all safely collected, it was imperative to continue my journey south towards Chile and on to Buenos Aires, some 4,185km away. But getting to the finish line by New Year's Eve was starting to look increasingly unrealistic.

After waving goodbye to Danica, her father and Maria, I got back behind the stroller. Me, that is, and an unexpected police escort through the streets of Chorillos with Danica and friends following in their car, all securing my exit from Lima and reuniting me once more with the Pan-American Highway. This convoy had assumed almost regal characteristics, but instead of the obscured windows and diplomatic number plates, it was a small, scruffy runner and a careworn stroller at its heart. Almost immediately a family pulled over to find out what was going on. They were clearly amazed when they heard my story and dug out a packet of biscuits for me while refilling all my water bottles. This small act of kindness is exactly what I needed to reignite my passion and set me on my way.

Danica had put me in contact with a host of people along the coast south of Lima and my first port of call was Punta Hermosa to meet Gaby. On reaching the turn-off, Gaby and her boyfriend were waiting, and he was all dressed up in his running gear to join me. Even more touching was the sight of the group of people who had congregated around a local restaurant to cheer our arrival, and further sustenance in the form of black clam, and other local delicacies, had been laid on for the afternoon. Despite the warm hospitality, I was determined to make my first running day a meaningful distance and said my goodbyes in good time to make it to San Bartolo. While conscious of the need to run over 47km a day for the next 89 days, it is also true that after five nights with people around, I was craving some time alone.

San Bartolo was a surf town that was clearly out of season. As I walked the streets with my stroller, I happened to meet the owner of a clifftop hotel that overlooked the bay below. The rooms were unaffordable for me, but after a while he offered me a hugely discounted rate that I couldn't refuse. Together, we wrestled my stroller around the narrow corridors to my room and then sat down to chat. This was a treat, as my host had rather extravagant opinions about the universe, and conversation ranged from his belief in UFOs to the trips he organised to experience extra-terrestrial life. On the road, you have to be prepared to hear views that differ from your own, and learning how to respectfully interact with the people you meet is as much a survival technique as finding food and water.

My target the next morning was Bujama, some 46km southeast. Wally had gone above and beyond any realms of generosity and had offered his beach house to me for the night. Not only that, but he was going to drive more than 80km to meet me there and let me in. I covered the distance as quickly as possible only stopping to refuel, mindful of not keeping Wally waiting. He had decided to run the last kilometre with me, so was waiting by the road when I got there. Despite wearing jeans and evidently very little recent training, Wally dug deep as we followed the beach and cut through the old rock tunnels that led to his magnificent house. I was treated to a tour of the house and the beach, despite some persistent rain. This even included an impromptu swim. At the suggestion, I gingerly looked out at the cold ocean and icy looking waves, but before I could register my reservations, Wally was running towards the sea in his underwear. Left with no option but to follow, soon there were two men splashing in the waves of a deserted beach. It was an invigorating moment.

Peruvian hospitality continued in the town of Cerro Azul, where I would meet Luz Maria. To spend as much time with my hosts as possible, I raced through my marathon distance. On arriving in Cerro Azul, I was greeted by a group of smiling ladies. Luz Maria was flanked by Fifi, Shari and Fiorella and they led me straight to Mario's pizza restaurant on the town square and watched me devour a Hawaiian pizza.

Luz Maria's house was located metres from the water's edge and, once there, the first task was to assign me somewhere to sleep that night. This group's positive energy immediately made me forget that day's running, and we soon set out on our own little adventure. Luz Maria, being an excellent host, had organised a tour of a local archaeological site. The town had originally been a fishing village and probably part of the Yauyos indigenous people before they were overthrown by the Incas, who built a stone fortress on the clifftops. One of the main points of interest is the stone wall built into the actual clifftop and that remains there today. Our tour was conducted by the archaeologists, who even took us back to their laboratory where a huge surprise was waiting.

One of the main findings at the site was mummified bodies from the Inca period, some 700 years before, including two recent discoveries they wanted to introduce in person. Full of anticipation, we were masked and gloved up and shown into a laboratory with a body that had remained hidden for over seven centuries. The excitement in the room was indescribable as we watched them peel back the leaves used to wrap and preserve the body. Before long we were gazing at the face of a human who had lived in an empire long since extinct. Teeth and black matted hair were still in place and you could see skin clinging to the folded arms. It was an incredibly special evening and I left with a few leaves to remember this incredible privilege.

Before being thrust back into the solitude of the Pan-American, I had to meet Beto, the husband of one of the ladies I had met in Lima. The distances I was running each day meant that I was going to leapfrog their home, so it was decided that he would meet me on the highway and return me to the same spot the next day. But there would be one unplanned encounter before that could happen.

Running through the darker desert sand under a grey, overcast sky was monotonous, so hearing a voice shout out in English on this exposed coastal road put me on high alert. The man who approached me was clearly drunk and potentially high as he stumbled along the side of the road. Barriers up, I cautiously decided to engage anyway to find out this guy's story. He was British and had served in the British Army for a short spell before falling into a life of crime and 20 years in jail. His family had disowned him and, feeling abandoned by the British Government, he had moved to Peru where he felt he could live without the difficulties of being back in the UK and potentially also have easier access to drugs. He pointed to a rundown hut on the beach that looked like nothing more than a disused building and invited me over for a drink. As soon as he started to tell me how Martians had visited him there, I quickly explained that I already had somewhere planned for the evening and needed to make tracks. Ignoring me, his narrative veered into the truly bizarre. It started with him asking if I could see the person sitting on his shoulder. I replied that I couldn't, which prompted him to shake his head quickly from side to side, exclaiming that he had just seen it. This was followed by warnings that the world was about to implode due to everything that we had mined from its core. He went on to tell me that he was in fact an alchemist and all he had to do was pick up a big stone and it would be transformed into gold, but he would then be unable to carry it because of the weight. The stories got more animated and fanciful, including serpent men rising from the water. I felt sorry for him and a bit helpless. The two plastic bottles of clear liquid he carried were obviously not full of water and he clearly needed a doctor, both mentally and physically, but here he was just left to deteriorate and drink himself to death. There was nothing I could do but continue on my way and leave him standing on the side of the road rambling to himself.

While processing this interaction, I came to a more built-up area where Beto would be waiting. Throughout my adventure, I have been very wary of accepting any kind of lift even if just to retreat the way I had come, but here I felt confident I wasn't cheating myself, as Beto had agreed to bring me back to where I had been picked up. It's baffling that the whole concept of cheating even crossed my mind. This wasn't a first attempt or race; at the end of the day who would care? The answer was clear, I would! I set out to do something and I wasn't going to compromise it in any way.

Beto arrived in his big white pick-up, in which we tore down the highway towards Chincha Alta, known for its historical importance during the colonial period and its role in the sugarcane industry, and then on to his family ranch a little further out of town. The property was stunning. The main house had been developed into a hotel while the family had relocated to what used to be the old administrative building, albeit still vast and imposing. Evidence of its past was everywhere, like the hatch next to the main door where the field workers used to collect their pay. Arranged around a courtyard, in this house everything was decorated impeccably. This was a family clearly used to entertaining as my room was a bunkhouse that I was to share with their daughter's boyfriend. Beto gave me a chance to clean up, before taking me on a tour of the property, showing me where he was planning to retire to when his son took over the main house. As it had been in Panama, cockfighting was a huge sport here, and the walled enclosure of the second house had its own cockfighting ring and over 200 fighting cockerels in training. Horses, llamas and party rooms completed the picture of this grand South American estate.

Having been beautifully cared for by Beto and returned to the highway, I set my sights on reaching Pisco the next day. The running was hard, and my motivation and momentum were flagging. Even on this relatively flat road, the slightest inclines wore me down. This was a very poor part of Peru and imagining the lives of the inhabitants of the small huts along the roadside contributed to my feeling of melancholy. But then, as had so often been the case when things took a downturn, I got the boost I needed.

My phone pinged to announce receipt of a donation to CALM, the charity I was raising money for. A few minutes later, another ping and then another. This unusual uptick in donations gave me a reason to take a rest and investigate what was happening. Scott, a friend and former colleague had put up a post on Facebook, detailing what I was doing. He had sensed that with everything else, I was unable to promote my charity angle sufficiently, so had decided to take matters into his own hands. Needless to say, this was a bighearted and selfless act, for which I remain truly grateful. Scott's awesome blog post finished with a link to Mark Zuckerberg and Stephen Fry, thanking them for sharing my post. The post alone had generated a few donations but what really kicked it into action was a tweet from one of his friends also asking for Stephen Fry's online assistance. '@Stephenfry, a very special guy needs your help, please click on the link below and read.' And read he did. The follow-up tweet from Stephen Fry began: 'Good gracious, what a guy...' These three messages generated nearly £1,000 for charity over the next few hours and raised my spirits no end. My ragged progress had transformed into jumps of joy every time I heard my phone ping. Scott had no idea I was struggling, but nonetheless, he gave me just the jolt I needed.

This was important, because the road was getting less hospitable. To get inland to the town of Ica, there was a tough challenge ahead, with over 450m of vertical ascent heading straight into the baking hot desert. The only alternative to making a straight dash of the next 65km was camping partway in a waterless desert, which wasn't really appealing. That said there was no denying that my body was tired before I even started.

My first stop was at a small house on the side of the road where an old lady was selling cake. I sat and chatted to her, in a blatant attempt to put off the inevitable running. She was completely blown away the concept of my run which gave me extra impetus to push on.

Piling the harshness of the desert onto the steeper gradient made the running hurt, both physically and mentally. What's more, among featureless dunes there was nothing here to inspire, so to keep going you needed to go into yourself and just push on. My only distraction was from the few travellers on motorbikes who stopped for a quick selfie before shooting off in the other direction. To the giant lorries ploughing past, a solitary runner was of no interest whatsoever.

Before I could enter Ica, there was a military checkpoint to contend with. After this formal introduction came the feeling of danger that heralds all bigger towns, and it was compounded by the feeling that more and more people were looking at me as I neared the centre. To counter this claustrophobia, I was drawn to Huacachina, a small village about 5km to the west of Ica.

Huacachina is a picturesque oasis village, known for its stunning lagoon and surrounding sand dunes. According to local legend, a beautiful princess named Huacachina created the lagoon from her tears after fleeing a young hunter who pursued her, ultimately transforming him into a sand dune. The myth is said to reflect themes of love, loss, and the connection between nature and humanity. Although diminutive, the village is a hugely popular tourist highlight for both Peruvian and foreign visitors. Due to human activities, unfortunately the water stopped seeping into the lagoon in the 1980s, requiring it to be artificially refilled. These efforts are ongoing.

The road to Huacachina climbed over the brow of a hill and when the lagoon came into sight, it was true that there was something magical about it. Nestled in huge sand dunes, the still water was surrounded by green palm trees, while a crowd of hostels and restaurants competed to entice the tourists.

While letting the beauty wash over me, I started to fear the worst for potential accommodation. This was clearly one of the most popular tourist sites in the area, and it was a public holiday. On the way into the village, sign after sign declared that the properties were fully occupied. I went from hostel to hostel pleading for a room or a space to camp but found nothing available. Occasionally a bed could be had, but with nowhere secure for my stroller, and after 76km of running through sunbaked sand, I was struggling to deal with this rejection. Anger and frustration began to creep in, neither helpful when trying to negotiate with busy hostel owners. As darkness closed in, I became more desperate and frantic. Taking a couple of minutes to calm myself, I remembered that sometimes extra expense on accommodation was unavoidable because recovery was the priority. In extreme circumstances, you have to be willing to use all the tools at your disposal. On a hotel booking website, the credit card was my last resort to get me out of trouble.

At my destination for the night, the concierge took pity on me with a small discount to make the blow more manageable and finally I was in a real bed, exhausted after one of the longest days of the expedition. The beauty of the oasis would have to wait for the next day.

After making the most of the buffet breakfast, I packed up and moved to the hostel next door where a vacancy in a dorm came up after the night's festivities were over. That morning would be devoted to recovering from the physical and mental strains of the previous day. By afternoon I was ready to hike up the vast sand dunes that tower over the oasis, where I sat on the ridge reflecting on events while watching the sun set over the boundless desert.

Despite being surrounded by tourists and backpackers in Huacachina, once again I felt a little alone. Luckily, that evening I found a group of foreign travellers who took me in. It was one of the girls' birthdays and we all sat around chatting until late to mark the occasion. Staying on to have more fun was admittedly tempting, but the angel on my shoulder gently reminded me of my target of Buenos Aires.

Leaving the green farmland surrounding Ica behind me, the next day and those that followed I climbed back up onto the desert pampa. This place was so desolate and empty that it was even devoid of the small villages I had come to rely on. As the wind's long fingers rifled the sand, I scoured the horizon for anything that would resemble shelter. At a disused building, some cautious checks ensured that no one had been living there recently. Satisfied that I wasn't trespassing or going to be interrupted in the night, I set about pitching my tent. However, the howling wind made the whole process too difficult and the baked ground rejected my pegs, confirming that I had to repack and push on. Camping in the open desert really wasn't an option and I would have to rely on the kindness of strangers.

At last I found an inhabited dwelling but was immediately turned away. Tail between my legs and as the sun started to retire behind the hills on the horizon in the west I pushed on, before finally finding another small settlement on the side of the road. At first it looked no more than a hovel, but there were sounds of life and I approached with a sense of desperation rather than the caution I should probably have adopted. Ominously, this holding was a collection of small huts made of debris and surrounded by the rubbish and used tyres that lay on the desert floor. It was a depressing place to be and a striking contrast from the places I had stayed closer to Lima, yet I was nevertheless overcome with relief when the owner agreed to let me camp. Unlike other hosts, he immediately disappeared and left me to get on with setting up. At a small parcel of flat ground, I found shelter from a nearby hut and could just about get the pegs in the ground, which was enough to ignore the rubbish piled up next to me. The desert was cold and uninviting that evening and once my tent was set up, I crawled inside and made some food before passing out.

If a long night's sleep was my goal, then I was to be disappointed when at five in the morning the resident animals awoke. Noah had nothing on this place. After trying to ignore the racket for as long as possible, I emerged to find turkeys, guinea fowl, ducks, chickens, you name it, all waddling around in pairs clicking, gobbling and quacking. This commotion had woken the dogs, who all started to bark. I hastily made some porridge, packed up and made my way back onto the road.

Qualitatively, this section of desert was so different from all the other similar environments I had been in. Fog hung thickly over the road and the air was ice cold. For the first time in months, I was wrapped up in everything I could get my hands on, and even the exertion of constant running did little to warm me. Motivation would have been low had it not been for the prospect of the magical Nazca Lines, only two days hence.

My day improved as the kilometres fell away and the sun burnt through the mist. The grey, lifeless desert slowly acquired beauty and with the arrival of hills, things got more interesting.

Some people are never happy, and by lunchtime, I was cursing the heat again, desperate to find some shelter. My chosen route was less suited to lorries and while more difficult, I enjoyed not having to jump to the side of the road every few minutes. Strangely, the more difficult the terrain, the easier it is mentally to deal with the task in hand, and this short stretch offered everything I needed to keep going. Steep ascents with crude tunnels blasted through the stone opened up to beautiful views of green valleys and roads that dramatically swept downhill. This variety made all the monotonous straight flat running worthwhile.

Palpa, while not the prettiest of towns, is archaeologically important and is home to the Lineas de Palpa, which some historians believe to be the possible precursor to the better-known Nazca Lines. By mid-afternoon I was completely wiped out from the terrain and while the lines were close by, I had no idea how to view them. With the Nazca Lines just a day's run away, I decided to just focus on getting there instead. It was yet another reminder of the sacrifices this trip required, as among rich historical and cultural resources, sometimes I simply had no capacity to explore them all.

The next morning, being filled with enthusiasm for the Nazca Lines was a delicious feeling. This excitement propelled me through the first section of the run along a valley with steep cliffs reaching up on either side. A pick-up pulled over and a group of workers got out for photographs with me and kindly donated some fruit. Further on, I crossed paths with a Canadian cyclist called Jerry who was cycling south on his way to do some rock climbing in the Andes. He had rough-camped the night before which made me feel a little guilty that I had not. But on learning that he had only been cycling for a few days and I was on day 426, I didn't punish myself too much.

Just before the Nazca Lines was a small museum celebrating the life of Maria Reiche, set in her former home. She was a German mathematician, geographer and astronomer who, in 1940, worked with Paul Kosok, the first westerner to study the Nazca Lines in any depth. Maria dedicated the rest of her life to studying them and when she died in 1998, was buried in Nazca with full honours. She spent not only time, but considerable amounts of money trying to ensure that the Nazca Lines would be protected and built observation towers along the Pan-American so people could view them without causing further damage.

These monuments are now a UNESCO world heritage site and to see them I climbed the final hill to reach the plain on which they are drawn. The lines were created by etching shallow patterns into the ground by pulling the white pebbles aside to expose the redder ground below in what is known as a geoglyph. The largest of these is up to 370m, but countless others scatter the plain. It is believed that they were created between 500 BC and 500 AD by the Nazca culture and have remained visible, preserved by the lack of rain or wind in this spot. The shapes include birds, fish, llamas and various animals.

My Nazca Lines experience was limited to an observatory on the side of the Pan-American, but from here I managed to see the Nazca Tree and the Nazca Hands. Although smaller than I had expected, both were still hugely inspiring. It is hard to comprehend how and why they were made and even today there is a huge debate about what their real purpose was. For me, their beauty and mystical nature is enough without needing to know the how or why.

Leaving the geoglyphs behind, the road across the Nazca plain was a hard slog and it was all I could do to focus on getting to Nazca. Over this arrow-straight 20km stretch, a punishing headwind fought me at every step.

At Nazca, part of me wanted to take a plane to view the lines properly but on grounds of budget and time, I decided to skip it. Also, while viewing the lines with strangers would be exciting, I wanted to share it with someone—a feeling I noted more and more as the adventure progressed.

By now I had been on the road for over fourteen months, but a few days stand out as being really tough, and leaving Nazca was one of them. Drained in body and soul, I felt bereft of motivation after only one day off in the last 11. Still, my brain urged me to keep pushing if I was going to have any hope of sticking to my schedule. On the outskirts of Nazca, a vain attempt to kickstart the motor meant eating junk food but that only weakened me. Strength gone, after 7km, I pulled into a huge service station at the side of the road. My options were returning to Nazca or taking some time now to reset the day. I had set out that morning with a big day in mind and I was damned if I was going to be defeated. Leaving the stroller on the forecourt, in the restaurant area I ate as much real food as I could.

After a good rest and stern talk with myself, I hit the road, but this time with renewed determination to run. The road wove through small communities with shops selling beautiful fruit that injected the extra energy I needed. This was a very good thing, as the road got much harder and I found myself battling uphill through a rocky red desert where the dust-filled wind lashed my face. With no established villages for over 60km, today's ambition was to cover as much distance as possible and find any respite from the relentless elements wherever it occurred. I can remember shouting at myself to continue, trying to convert tears into pure adrenalin. After nearly forty kilometres of hardship, I spotted a small roadside restaurant ahead and prayed the door would open for me.

At the small junction in the road, this turned out to be a sad-looking collection of cabins clustered together. One was the Restaurant Parinacochas, comprising a couple of brick huts sheltered by reed matting and a shaky roof supported by thin wooden timbers. I asked the two older ladies in one of the huts about eating and sleeping there. Showing no emotion, they agreed and showed me through the hut to a flat area behind the restaurant. It was clearly where they dumped unwanted debris, but there was a flat bit of ground and reed fencing to provide some shelter from the driving wind. This was a funny answer to my prayers, but at this point I was grateful for anything and went about driving my tent poles into the rock-hard ground. My dinner was some deep-fried chicken (with added fried flies for crunch) and rice. It looked so unappetising, but it was fuel and I didn't have to cook it. Trying not to think too much about what I was eating, I filled my stomach and retired to my tent. Even though that morning I had been struggling to cover a kilometre, somehow I had managed to press on and find food and shelter in the middle of nowhere. The discomfort of the day only made it more rewarding. I snuggled up in my sleeping bag, opened my laptop and watched The Secret Life of Walter Mitty to immerse myself in fantastical adventure and distract me from the real challenges I faced.

Over the last couple of days the wind and exposure had been getting to me. To make things harder, I was pushing the additional weight of water in my stroller that I needed for the long stretches between civilisation. Thankfully, the route to Lomas was only 44km and would be slightly easier by being mostly downhill. With spirits low, I decided to dip into the care package my mum had sent to Lima for a feast of French pate on Ritz biscuits and guindillas.

Lomas was little more than another collection of restaurants clustered around a junction. There was a hotel advertised 5km towards the coast but in total this was a 10km deviation, so I opted for the convenience of the location and camped under some palm trees.

Between here and Yauca was 40km of road interrupted by sketchy-looking settlements. The very thought of stopping at one only spurred me on. Perhaps I should have, as just before arriving in Yauca I was pulled over by the police for a casual interrogation. This started with all the usual questions, but this time it progressed to more personal questions about my past life. The more senior of the officers started asking how much money I earned and what I had in the bank. Affronted, I refused to answer him and instead turned the questions back on him, asking how much he earned. This wasn't a wise move but I was tired and irritable. He too refused to answer although luckily it shut him up and he left me to continue.

Of all the things I was expecting from Yauca, what I saw was not it. The town is set in the bottom of a river valley and on either bank stand hundreds of olive trees. As I got closer to the town, unattended olive oil stands cropped up. In search of somewhere to stay, I surveyed the makeshift brick houses lining the road and found very little welcoming about them. The hostel I selected was crumbling and everything had been cobbled together. Despite boasting a 'Presidential' sign, my room was little more than a basic bed and a light, while the shared shower was a hose protruding from the wall with a functional tap and no screen. There was no chance of getting my stroller up the tight staircase, so it was relegated to a gated carpark across the road, which caused me no end of worry. I had stayed at many hostels in similar states of disrepair but this one had its own charm, mostly due to the manager who lived in a small one-room apartment that more resembled a cell. He rushed around, proudly showing me the amenities, making me feel lucky to have the best room, all the while trying to reassure me that my stroller would be fine across the road.

The next morning I woke up early to set off down the Pan-American Highway, only to find the road closed due to a rally race. The police said, quite understandably, that the route was closed even to those on foot. I had no choice but to book back into the hostel and wait out the day. The race itself wasn't particularly high-end but joining the spectators was fun. When the cars weren't roaring past, as the only gringo in town, I became the crowd's entertainment and enjoyed laughing and joking with them as they became increasingly inebriated throughout the day. That evening I ate in a small restaurant next to an extremely drunk Peruvian chap who found it hilarious to discuss his sex life in graphic detail. He was the only one laughing.

Happy to be set free, the olive groves receded as I swung down to the flat coastal road where the Pacific crashed onto an empty beach. The road passed through Tanaka, a coastal town that looked like it had been deserted halfway through construction. Half-built houses skulked against larger expensive-looking condos. The road then climbed into the mountains, following several switchbacks as it avoided large, almost-lunar boulders. Then, with an onrush of emotion, I was reminded of Scotland on entering a section with bright green grass covering the hills. At one point, it even started to rain, which was the first time in a long time I had to get the waterproofs out for the long sweep back down to the coast. The prospect of Chala down the road seemed uninviting at best in this drizzle.

At the foot of the hill I heard barking and turned just in time to see a huge guard dog striding across the parking lot of a local establishment. Having learnt that running is not the best option in these situations, I turned my stroller to face the approaching animal and spotted a short, plump lady, dressed rather provocatively, burst out of the windowless bar to shout at the dog. When it saw I posed no threat, the dog slowly returned to the lady who was now shouting apologies at me instead.

Chala, not helped by the weather, was drab and downbeat. The main street was made up of the usual selection of food stalls and shops selling whatever they could. The bus station was bustling with people coming and going past a sad amount of others who were merely begging. Clearly, nothing in this town was going to lift my spirits, so I resorted to finding a hotel and booking in for the night. On requesting a room that would fit my stroller, I was shown the biggest one only to realise there was no hot water, no internet and no power due to a power cut.

It was in Chala that I hit a real low point, something that rarely happened but seemed to be happening more frequently. Together with the grimness of this town, the distances I had been running, the lack of rest, the harsh environment and the weather were taking their toll. Next morning I awoke with absolutely no motivation and all I wanted to do was lie in bed. So that is what I did. I had been fighting day after day and had lost the will to continue that fight. My biggest challenge that day was convincing myself that I was allowed to feel defeated every once in a while. With the eyes of so many on you, it is very easy to convince yourself that because you are continually updating people about your progress that you will be judged if you stop. The reality, and what I tried to tell myself, is people were not watching that closely and those who were, were willing me on. Getting my mental game back onside was my priority and I spent the day doing little else.

CHAPTER 29 - MICROSOFT WANTS A MEETING

DISTANCE TO BUENOS AIRES: 3,533KM

Luckily my day off had worked, and I woke feeling motivated to run. The more days I took off, the harder it was going to be for me to arrive in Buenos Aires on time and my daily target had by now crept up to 49km every day for 72 days without a rest. While this seemed just about possible, another episode of fatigue, injury or logistical failure could put a spanner in the works.

Camana was the next major town about 220km away and I knew that over the next four to five days of running there would be very little in the way of civilisation. On the way out of Chala, it was frustrating to observe that the old town was quite pretty whereas I had spent my day off a stone's throw from it wallowing in self-pity. Nothing could be done about that now, and after 13km up and over a hill the road started to follow the coastline into extraordinary scenery. A small restaurant was advertised by the side of the road, where I stopped for more fried chicken and chips. While attempting to enjoy the respite from the elements, I had to suffer the constant conversation from the owner and his countless questions. Perhaps on any other day I would have felt more tolerant, but my heart just wasn't in it at that moment. That night I pitched my tent just off the road, behind a hump and on top of a cliff overlooking the ocean and, from this vantage point, just being back in the tent was helping to get me enthused again.

Along this rugged coastline, cliffs rose to the left of the road while the Pacific Ocean beat the rocky shore below on my right. The day's plan was to run the 40km to Atico, but as I arrived there at midday it seemed prudent to push on and cover some extra distance in the afternoon. The distance between the mountains and the ocean closed considerably and before long the road had become a mere ledge cut into the cliffside, with just enough room for the traffic and very little spare for a person and a running stroller. While the running here was perilous, the scenic rewards more than made up for it, with dramatic drops into the ocean and a pod of dolphins seeming to cheer me along the coastline. With these amazing distractions, I managed to eke out another 25km before it started to get dark.

I possibly could have foreseen that trying to find somewhere to camp might prove more difficult than normal, with nothing but sheer climbs and drops on either side of the tarmac. Finally, a small gorge that cut into the side of the cliff provided enough shelter and cover for a night's sleep. But as I had found even in North America, the only downside to such refuges is that they are the only places where cars can pull in and as a result, they become toilets, with piles of excrement and paper towels strewn around. My challenge was to find a patch of ground that was clean enough to pitch camp for the night. As the sun set, to avoid spending longer in that area than absolutely necessary, I ventured back out to the road and watched dolphins as they fished in the waters below.

After swiftly packing my things the next morning, I had to climb away from the shore and run a rollercoaster of a road that made it hard to get into a groove. Simply to get the energy to keep going, my best bet was to take a break, so at about 20km I stopped at a small restaurant and ordered some fish and chips. After a ridiculously long wait, I noticed the lady coming back to the restaurant with bags and realised that she had had to go and buy the ingredients for my meal. Inadvertently this had done me a favour and given me a chance to rest up before setting off for a longer afternoon stint.

The next small town was industrial with a big refinery dominating the port. Perhaps, given the effort this day was requiring of me, I could be forgiven for feeling a real hankering for chocolate and seeking out a shop. It took me about five minutes to choose from the meagre selection they had on offer before getting back on the road, determined to make it to Ocoña that evening.

This unremarkable experience was swiftly followed by one that was far more noteworthy when I heard a deafening crack followed by a crash. Looking around to see what had caused the noise, I rounded the next corner to see clouds of dust rising from the road ahead. As I got closer, I could make out through the haze that a landslide had blocked the road at its narrowest point between the cliffs and a large rock. This was my second landslide of the adventure, after being forced to make a detour in Mexico, some months previously.

The traffic started to back up but luckily no one seemed to have been injured by the rockfall. Being smaller and more nimble than the cars, I was able to weave between the traffic and get to the blockage only to find it impassable. While the damage was considerable, I worked out that with a little help I might be able to navigate my way across the rubble and enlisted one of the nearby drivers to help carry my stroller. By the time I got to the other side, all the drivers were clubbing together to clear a route for the smaller vehicles, so of course I lent a hand. Before long we had managed to clear a small passage through that at least allowed some of the traffic to start moving.

Feeling I had done all I could, I pressed on suffused with adrenalin. As I ran on, it dawned on me that if I hadn't stopped for chocolate then there was every possibility that I could have been running directly below where the rockfall occurred. My mind started running through all the different narrow escapes of this journey, which was enough to distract me from the steep climbs and long descent into Ocoña.

Ocoña is a small, quiet town that looks like it had once tried to entice tourists but has since forgotten about that. My small hostel was perfectly adequate but very little was going on in the main town, although I did discover that fresh river prawns from the Ocoña River were a particular delicacy. Needing no further incitement, I went on a mission to find some for dinner. Just before the bridge I had crossed earlier in the day was a cluster of restaurants that advertised the local speciality. The prawn soup, while delicious, did not quite fill the hole created by over 50km of running, so I supplemented my meal with some local bread from a nearby hut.

From here, it was 57km to Camana and I was set on getting there in one stint. Camana had a strange pull for me because it was where British adventurer, Ed Stafford, started his trek along the course of the Amazon in 2008. I had been reading his book and was filled with respect for what he had achieved, so thought it would be fun to stay in the same hotel he stayed in. The issue was how to get there. Running into Camana was proving hard on my body as I had been averaging about 55km a day for the last four days and could feel it taking its toll. The only thing that could replenish my resolve was the epic scenery, as the desert sands blanketed the land, only punctuated by gaping canyons. The road drew a hypnotic black line between the beach and the desert, under permanent threat of choking under the fringes of sand that built up on either side. This was like nothing I could ever have imagined before setting off on this adventure, and I am grateful to this day for the experience.

As I got closer to Camana, the weather changed, and once again it summoned my rain jacket. As the drops fell more heavily, I opted to shelter in a small roadside restaurant and ate well before taking on the hustle and bustle of Camana itself. The restaurant I found was, like most restaurants in this part of Peru, run by a family and felt more like I had been invited into someone's house for some home-cooked comfort food.

Using the relevant chapter in Ed's book, I was pretty confident that I'd found the hotel he stayed in based on the description given. The actual name was not disclosed, but everything he had written indicated the nicest hotel in town, the Hotel de Turistas. I decided that was a good enough reason to check in and was shown to a spacious room on the ground floor.

My motivation problems were not going away and despite running 220km over the last four days, my usual get-up-and-go was rapidly dwindling. Concerned, I decided to take yet another day of rest to knock this negativity on its head. In this case, it meant a day walking around the town, eating as much as possible to refill my reserves. Looking back on my notes, it makes me smile now that I wrote 'getting fat from bad diet.' After running 13,700km, how I looked was certainly not something I needed to be worrying about. What was certain was that a change in diet was needed as I was clearly not eating efficiently.

During a long conversation with my girlfriend, whom I hadn't seen for over 100 days, I encouraged her to book her ticket to Buenos Aires for New Year's Eve. It wasn't an easy sell as there was still 3,300km to run with just 66 days to cover the distance. She wanted reassurance that I would be able to run 50km a day on average and not falter, which was awkward, as that was the very thing I wasn't 100 percent sure I would be able to do. She suggested that if I didn't make it on time, then the only solution would be to get a lift to Buenos Aires, spend New Year's Eve with her and then return to continue my journey. I reluctantly gave my word that I would make this concession if I hadn't made it, but inside this just fired up a new determination. Taking a bus into Buenos Aires would represent capitulation, and no matter how hard I had to push myself, I would be running in triumphantly. Perhaps, on reflection, she understood this, and I have her to thank for reigniting my inspiration and motivation.

That night my conclusion was that I needed to add an even more challenging element to the journey to give me a sense of urgency. My belief that I would be able to achieve it had faltered due to longer than expected delays in Quito, Lima and other towns. But as I sat in Camana it suddenly felt possible again; it had to be. The goal was an average of 50km a day, every day without fail and just to make it more difficult, I would have to cross 700km of the Atacama Desert and then navigate altitudes of over 4,800m in the Andes.

The next morning, I received another call from my girlfriend who had found out some distressing news from her home, to which I immediately said I would stay in Camana another day so I could be there for her. My new drive to reach Buenos Aires had not got off to a good start and I questioned myself, rightly or wrongly, about whether I was using this situation to justify another night away from the road. Life events are unstoppable, however, even when you're thousands of miles away from the people you love.

After two days of rest, I was all packed up and making my way to the town limits. I knew that the running ahead was going to be tough, with a 1,300m climb up to the desert plateau. Although the initial 10km was nice and flat along the coastal road and past hotels and hostels, any complacency was swiftly erased by the 17km climb on a road that wound through exposed valleys straight out of scenes from Star Wars. A dense fog set in that brought visibility down to a mere 5m and made running on the tight road more than a little fraught. In these conditions, cars and lorries had to be detected aurally, as it was impossible to rely on just seeing them.

Finally, the fog receded in favour of a vast flat area that stretched as far as I could see. This monotony was not much better, and I struggled on another 20km, determined to hit my planned 50km a day before identifying potential sleeping spots. With no cover, the least visible place was behind a large white sign just off the highway, and while this would not shelter me, it made me feel slightly more secure. That night I sat and observed one of the most amazing sunsets I had seen on this adventure, with the huge sky kaleidoscopically changing as the sun sank below the horizon. Being back in the desert and camping out was certainly aiding the turnaround in my mood.

A jaunty 24km uphill greeted me the next morning as my desert running resumed with the soundtrack of Bob Dylan on my speaker. On both sides of the road, farmers had reclaimed patches of land and the greenery jarred in such a desolate environment. The road then sharply descended into a huge gorge that split the landscape. In what was presumably a reasonably common occurrence, a large lorry had misjudged the road on one bend and lay a few hundred metres down a steep cliff, highlighting just how dangerous these roads were. A concerned crowd of people had gathered to watch as the recovery got underway.

As I climbed out the other side of the gorge my phone rang. With lungs still screaming from the ascent, I answered to find myself speaking with the marketing team from Microsoft's London-based agency, 1000heads. They wanted to know if I had time to jump on a conference call. Of all the places I had held work calls, this was the most bizarre. The team had recently done some brainstorming for a new Microsoft campaign entitled, 'Do great things' and wanted to know if I would be interested in being part of it. Of course, it was an amazing opportunity and I jumped at the chance. They explained that a film crew would therefore come and join me to document part of my journey. My spirits soared and I spent the next few kilometres shouting for joy into the emptiness. The final details would be made over the course of the next few days.

To complete my day's target, I still had 20km to run and the long, straight road was lined with small crosses where people had died due to crashes. Not wanting to become one of them, I stopped at a small petrol station with a restaurant in the middle of nowhere to eat something decent. This restaurant was exactly how you would imagine eateries in an old western film. It should have had 'Saloon' painted in art nouveau lettering on the door to complete the effect. An old lady shuffled around serving the few guests who came in from a set menu of a watery soup, followed by meat, beans and rice accompanied by a cold drink. I eked out as much time inside as possible, trying to avoid the midday heat.

As the day came to an end I descended into a gorge, running through long tunnels, past dusty mines and many more memorials crowded on the hard shoulder, until I reached the village of Vitor. The place was nondescript and the lack of any accommodation made it clear that very few people took the time to visit. That said, it nestled in a canyon with staggeringly beautiful cliffs and gorges of green vegetation running along either side of the river that coursed through the middle. At one end stood Ampato, a dormant stratovolcano that soared 6,288m and dominated the horizon.

The local police allowed me to pitch my tent on the village square and assured me they would keep an eye out for any potential problems with the locals, who were already gathering around, curious about this stranger making camp.

Barking dogs throughout the night and a constant fear that someone might just try their luck ensured a restless night, and I was glad to pack up early. After a couple of greasy egg sandwiches, my run up the other side of the canyon commenced. The road had been blasted through the rock, creating a walled corridor about 20m high on each side and it continued to climb to about 1,700m.

By lunchtime I was on the outskirts of La Reparticion, where I was surprised to find vines growing in neat lines on one side of the road and fearsome nopal cactus on the other. Still fascinated by my surroundings, my stomach was alas not feeling right and breakfast was coming back to haunt me. I picked up the pace, frantically scanning the side of the road for a restaurant or service station with facilities. It wasn't until I reached the centre of town that I finally found somewhere and was so desperate that I abandoned my stroller on the forecourt and ran for the toilet, making it just in time. As horrendous as this was, I had only covered 20km that day and I was not going to let a bout of diarrhoea stop me from hitting my 50km daily target. It hadn't let me down before, so I opted again for the Coca-Cola remedy to try and kill the bacteria and give me the energy to carry on. But only about 5km further on, I was dashing off in search of another abysmal petrol station cubicle.

The road out of La Reparticion was through a flat dirty desert, strewn with hundreds of small one-room huts. In the oppressive heat it took everything I could summon just to keep going, but my energy seeped away with every stride. A brand-new service station shining in the distance provided a brief reprieve and as I entered cool air from the hardworking air conditioning gave me succour. Even small changes can drastically lift my mood, so I took as long as I could in the aisles picking out cold cans of juice, ice creams, anything that would cool my core down.

Air-conditioning is an incredible thing to come into but is hellish to leave. The heat outside felt even more oppressive on leaving, and to make matters worse I had no idea where I was going to end up that night. My available remedies for the immediate issues were to take on as much water and food as possible and get back into a rhythm. Sleeping arrangements could be the next task. The road continued in a straight line for another 12km before veering to the right and running along the perimeter of the Mariano Melgar Airforce Base. Having somehow clocked up 50km, I turned my focus to finding somewhere to camp. The military base was a problem, as I knew camping in the vicinity would be a strict no-no.

The road stretched as far I could see, aiming directly at a range of mountains in the distance. I couldn't determine exactly how far away they were, but I knew that it was a lot further than I would be able to cover in the time I had left before dusk. Then came the fairly unequivocal sign: 'Zona Aeronautica – Peligro de explosión' with an image of a fighter jet dropping bombs. The only interpretation was that this was where the Peruvian Air Force conducted their training and I was about to run straight into it. Turning back wasn't practical, so the only remedy was to find the best spot to camp and hope for the best. To my right the sun was setting, turning the sky into the most spectacular array of changing lights. Clouds were forming in the distance and to my tired mind it felt like the mountains were holding them back. With no other option, I settled on a cluster of disused buildings for the night. When the traffic was clear, I ran directly there and hid from the road as best I could. My choice of campsite begged some questions as the walls of these buildings were peppered with round holes that could only have been created by one thing—bullets. Just a few kilometres away, jet planes were audibly taking off. However, even the fear of military attack wasn't enough to prevent me from drifting off after running 62km with crippling diarrhoea.

Strangely unperturbed by dreams of warfare and military manoeuvres, I woke up feeling very refreshed. The mountains on the horizon were calling me, but rather than climb them, I weaved my way through before the daylong descent into a valley that ran out to the Pacific. The Tambo River provided the local restaurants with more freshwater prawns, as had the Ocoña at the last estuary, so it seemed only polite to stop at one of them to refuel and calculate how to navigate the next desolate stretch on my dwindling supplies. According to my father's calculations, there was a diversion ahead that could cut my distance while keeping me closer to small villages and avoid a huge desert with absolutely nothing for miles and miles. It was another reminder that, even though I was out here alone, when adventuring there are people back home who are feeling every step, every risk, and every joy.

The road I took cut west to the coast and then followed it south to the Chilean border. Away from the main road, the pace of life seemed to slow and it was bliss to escape the constant battles with large lorries and speeding trucks. Chewing on sugarcane as I ran, my new surroundings seemed to embrace me. Onions seemed to be the main crop and the harvest looked brutal for the surprising number of older people who tended to the fields, especially as there were no tractors or heavy machinery to help. That said, the workers were still able to muster a smile and a wave as I ran by.

When I arrived at the town of Cocachacra I started to worry that I might have been wrong to take this deviation. Police officers dressed in full riot gear lined the streets, holding batons with the visors down on their helmets. None of them seemed to take any notice of me and it wasn't clear what would warrant this heavy police presence in a town and area that seemed so docile and poor. Then signs started to proliferate on buildings, in shops and on homes. 'Si Agro No Mina', which translated means, 'Yes Farming No Mine'. The potential for a huge copper mine to consume the local area and destroy the agricultural community had been causing tensions to run high for some months, and the violent conflicts would last for years, until the eventual announcement in 2024 that the mine would open despite this spirited resistance.

That night I made it to the bustling town of Punta de Bombón, where blue and yellow tuk-tuks buzzed around the streets and vendors shouted to me as I passed. At my small hotel, it was a relief to get clean before venturing out to explore. Aware that my time in Peru was coming to an end, I was single-minded in my focus on eating as much of the wonderful local food, such as lomo saltado and ceviche, while it was still on the menu. Every traveller experiences a country differently, and for me, the cuisine is integral to really participating in a culture.

I was not the only one that was falling apart from the hardships of continual running. My stroller was definitely showing some wear and tear, as the ball bearings were starting to seize up and the spokes to weaken under the increased weight of the stroller. Not only this, but it had been over 4,500km since I had treated my trusty wagon to new wheels. After breakfast, I found a small workshop and tried to explain what the issues were. The mechanic merrily set about stripping back the wheels and managed to pull apart all the fittings and clean them with petrol and put everything back together. After all the work, I wasn't convinced that the wheels were in any better working condition, but at least imminent issues were less likely.

The first 20km after Punto de Bombón of the run lulled me into a false sense of ease as it ran in a straight line, parallel to the coast. Small huts dotted the hard shoulder with no signs of being lived in, and I mulled for a time what might have caused the former inhabitants to build them and then leave. My comfort was short-lived, as the road then climbed up into the hills that hugged the coastline, where the only shelter occurred when the road had been cut through a small rise, or from the shadow of a cactus. Back to dry rations, lunch was a packet of instant noodles eaten on the side of the road. Possible sleeping spots were limited so, at the 50km mark, a small cluster of buildings made of reed matting was a welcome sight. Even more welcome was the permission one of the men gave me to camp just off the main road.

Inevitably, spending so long on the road meant spending my birthday in transit. One year ago, on 1 November I had been in San Diego enjoying Halloween parties in nightclubs, drinking beer and eating mouthwatering American junk food. This year, the day started with crawling out of my tent, following my same familiar packing-up routine and eating porridge out of a small paper packet, but on this occasion, I did sing Happy Birthday to myself. While this all may sound very depressing, it brought me a lot of joy to be doing what I loved most in the world in the most extraordinarily basic place. I was indeed happy and that is what is most important.

Nonetheless, I still wanted to celebrate my birthday properly, even if it would be alone. The next town was called Ilo and was just under 50km away. The surroundings were changing as the desert sand took on a deeper, richer orange tone and the ocean once again lapped the shore to my right. The biggest distraction was a copper smelting plant that, while ugly, gave a welcome distraction to my running. This plant features in a Canadian documentary called The Corporation, which sought to highlight environmental issues caused by large industry, among other harms. The copper is transported there by train and then processed before being loaded onto large cargo ships for use in lucrative telecoms, tech and battery applications. On my way past, I noticed two ladies walking along the train tracks. At first, I couldn't work out what they were doing but a little chat with them revealed that they were picking up fallen copper from the trains to sell in town, a huge amount of work for very little return.

Ilo was another example of a town that had clearly seen more prosperous times. In the 19th century, a pier was built and Ilo became an important stopping place for trade running from the East to the West of the United States. However, with the arrival of the Panama Canal and the Transcontinental Railroad, the route though Ilo became all but redundant and the town began to suffer economically. Today its economy relies mostly on fishing and mining, so there is a palpable divide as the remaining rich mingle with the poor. Private members' clubs are advertised next to areas that are very run down. Arriving the day after Day of the Creole, a huge public holiday in Peru, meant that drunk partygoers still wandered the streets while music blared from huge sound systems. For once, I passed through unnoticed on my way to the centre of town. My only wish for my birthday was to eat as much meat as possible and after some local advice, I jumped into a taxi asking to be taken to the best steak restaurant in town. That night I ate like a carnivore king.

With the realisation that I may have overindulged the night before, I rang my girlfriend looking for justification for a day's rest, especially having run more than 300km over the last six days. Unsurprisingly, my pleadings fell on deaf ears and she informed me that she had bought her tickets to Buenos Aires so I'd better get running, birthday or no birthday.

Some 52km later, I found myself having a debate with an empanada vendor and petrol attendant about where the best place to camp was. It was decided that I pitch on a small spot on the top of a cliff overlooking a large area of wetlands below. I snacked on food from vendors at Ite while watching flamingos feeding in the shallow waters, streaking them pink, black and white as they took off.

My last night in real Peru was a small, semi-deserted village with the buildings mostly boarded up for the off-season. When I found a hostel, it was hard to say whether the owner was grateful to have a customer or annoyed to have to work. It was so hot during the day that doing anything outside was unbearable, forcing shelter indoors and rest for the big push to Chile and looming wasteland of the Atacama Desert.

One of my lasting memories of Peru will always be just how kind people were and my last day did not disappoint. A bus overtook me while running along a quiet road. It braked abruptly and the driver jumped out, jogging towards me to pass me a large bag containing chunks of fresh watermelon. Nothing was expected in return, this was just a Peruvian being a Peruvian. While I believed this kind of thing only happened in the wild places I had run, I am relieved that I have received similar acts of kindness while adventuring in the UK.

The final kilometres in Peru could not have been more fitting. Knowing that the border was by now quite close, I coasted through the open desert with the sun slowly setting on my right. Finally coming to a halt, I approached some ladies who were working the food stalls on the Peruvian side of the border to ask where they thought would be the best place to camp. They referred me to the military who were based in the old border buildings, which looked large and mostly derelict. The only evidence of life was music playing from a radio somewhere inside. I gingerly walked along the empty corridors trying to catch someone's attention, where a guard finally found me. After listening to my request, he disappeared to get the opinion of his superior officer. It seemed a lot of people were involved for what was really quite a simple request! Before long, I was being gently but firmly ushered from the building and towards a small patch of grass for the night.

CHAPTER 30 - CROSSING THE ATACAMA DESERT

DISTANCE TO BUENOS AIRES: 2,797KM

The stillness of the next morning was abruptly shattered by music blasting from the guardroom in the dilapidated border control offices. Now irreparably awake, I cooked my porridge, packed up my tent and proceeded to the border control to officially enter Chile.

From here, it was only 18km to Arica and the terrain was even more barren, with the rather worrying addition of signs warning of minefields, a haunting reminder of past conflicts.

My research about Arica itself had been limited, so I was frustrated to find that the only hostel it had uncovered was closed due to being out of season. With no Chilean SIM Card, I aimlessly wandered the streets looking for Wi-Fi and finally found a homely hostel owned by a retired New Zealander, so checked in for the night. One of the things I missed the most while on the road was the cinema and with Skyfall recently released, I was determined to find a cinema in town that would show it in English. Befriending a fellow adventurer, Chase, an Australian policeman who had spent a few weeks cycling around Bolivia, we resolved to watch some James Bond action together. A mutual lack of Spanish meant that we arrived at the right cinema but for the wrong showing, allowing us to feast on burgers while sharing our tales of adventure.

Sharing stories is never wasted effort because no two adventures are the same. It was now nearly 14 months and despite the various tricky experiences that I had had on the road, still Chase's stories of sleeping in disused houses, the rugged tracks and the gnarly terrain blew me away. Elated with this newfound camaraderie, the icing on the cake was to finally get our Bond fix, before returning to our hostel in great cheer.

It was time to take stock. The last 10 days had chalked up an average of over 50km a day. The solution was clear and I decided before tackling the Atacama, it seemed reasonable to take a rest day in Arica.

Ever since I conceived this expedition the Atacama and the Andes were the obstacles that people told me would prevent me from completing my challenge. For this reason, both had been ever-present in my mind throughout my journey and every time things got difficult, my mantra was that if I couldn't complete that section comfortably then the Atacama or the Andes would finish me anyway. It was a remarkably persuasive strategy. Most people thought me insane for even contemplating such a run, but having already run over 14,000km, I felt ready and even excited rather than scared. I had become accustomed to the distances, so these new challenges provided something to distract my mind, which was the part of me that was suffering most. If I could mentally get over a mountain, or negotiate with hostile officials, then getting my body and my stroller to do likewise was nothing in comparison.

The next morning brought a new level of unknown. On my way out of Arica, the modern structures of the town were succeeded by basic houses and slowly the desert swallowed signs of civilisation once again. The last image I had of Arica was of the huge Coca-Cola sign etched into the side of a mountain as a group of four-by-four trucks raced up and down the steep sides of the hills.

My first section was a long uphill that brought me out onto a limitless, arid plateau. A small hut on the side of the road was the last place to buy water for a while so I stopped for a cold drink while I had the chance. The further I ran the more the desert landscape engulfed me, suffocating me in its pulsing heat. Wherever there was any cover at all, I would take the opportunity to rest and cool down before striking out to face the fiery sun once more. Naturally, the closer to midday it got, the hotter it got and the less shade there was, so tactically my aim was to keep running to try and benefit from the wind that swept across the road.

Naively, I thought that this desert would be flat but that couldn't have been further from the truth. The road alternated up and down through this dramatic golden scenery. As the light started to fade, I had covered 60km and my priorities switched to finding somewhere to camp. Even here, a stand had been set up on the side of the road and, fortified by a small melon purchased there, I carried on climbing the side of a steep valley. It was exhausting, as every nook and cranny had something wrong with it until I finally found a perfect spot—a slip road that had been blocked by fly-tippers. But it was foul with excrement in the manner of similarly secluded places I'd camped previously. There was nothing I could do about that at this hour and in this location. Drained from the day's running and fighting a persistent wind I somehow got my tent into the hard, sun-baked soil and cooked some food.

Despite the surrounding unpleasantness, being treated to an amazing sunrise put me in a great mood. After breakfast I switched on Peanut Butter Jelly by Galantis (my unofficial soundtrack to this adventure) and danced my way through the packing-up routine. Glad to get going, the climb back up continued onto the exposed plateau, where a strong headwind swiftly eliminated any benefits of the downhill that followed. But I was rewarded nonetheless when I got to the small, quirky village of Cuya at the base of the valley. It resembled something from old cowboy movie sets, with small shops and restaurants hugging the streets on either side of the police control post. Once again, I took advantage of the restaurants and spent as much time in the shade as possible while enjoying a hearty meal.

After lunch, I ran along the floor of a long valley with high walls and a fierce afternoon wind that blew me along, filling my body with adrenalin. Any river that had carved this gorge had long since dried up, or perhaps only flowed maybe once or twice a year. By the end of the second day in the Atacama, I had covered nearly 130km and conveniently emerged from the rocky valley floor into a lush green area next to a river. The long grass provided perfect cover for my tent and I fought my way through the dense underground to find a small clearing.

The next day, the desert descended into steep valleys only for the road to pull me back up onto the plateau. With nothing to look at and nowhere to stop, I let myself be consumed by the rhythmic pace of the day, meditatively keeping my head down for another 60km while singing along to the Cat Stevens album I remember from childhood holidays. The problem of finding somewhere to pitch a tent re-emerged, as the ground had been so baked by the intense heat of the sun that even breaking the surface was a test. The best spot on offer was next to a parked lorry outside a restaurant, where I pitched my tent as an army convoy sped past and off into the desert. The military presence in northern Chile is down to a border disagreement with Bolivia after a treaty in 1904 rescinded Bolivia's access to the coast, making it a landlocked country. While it never seems to escalate into violence, old wounds run deep, and even in 2018 the UN was asked to intervene, so the militaristic feel is not likely to go away soon.

Less military in feel was Pozo Almonte, a curious town that retained some of its past charms. This part of the Atacama was once a mecca for saltpetre mining. Saltpetre or sodium nitrate was the main ingredient for explosives and fertiliser, and this part of Chile used to be the largest producer in the world. Old-world grandeur was evident in the long wood-covered sidewalks lined with stores and hotels, but decay was always present. On my way into town, I passed a lady who was so drunk that she stopped, pulled down her skirt and pants and urinated on the pavement. No one around seemed to bat an eyelid. Later, I noticed a few men sitting around sniffing liquids out of a can and smoking drugs from small pipes. There was no reason for them even to attempt to hide these activities. These were just a few signs of the crushing poverty in this part of Chile.

Despite this being a town, the best place to resupply for the long stretch ahead was the local service station on the edge of town and I was told that there would be few other opportunities to restock before Calama, some 320km away. With a heavier than normal stroller, brimming with food and water, the oppressive heat of the Atacama waited for me to resume this journey. The only way to survive was to split the day into 10km sections, finding any sort of shade I could, be it under a tree or crouched beside my stroller. My map showed a potential settlement at about the 60km mark, so I decided to push on, always preferring to stay where I could refill my water. On this occasion, I was pleasantly surprised to find a couple of restaurants and even a hostel located just off the Pan-American Highway. None of this was specifically aimed at me, and the huge dining room was clearly designed for the lorry drivers who also travelled the Highway. The food may have been basic, but it was plentiful and just what a hungry, parched runner needed. Even better, just across the street was a small, basic hostel and a bed for the night.

Even deeper into the desert was the Reserva Nacional Pampa del Tamarugal, a vast area of scrubby shrubs that lined the road before the desert once again took control. In the east, the outline of the Andes soared up into the sky, waiting for me at the end of this desert section and seeming to mock my human fragility. And so the Atacama rolled on, with old mines dotted along the roadside interspersed with new mines, where large lorries were loaded full of sodium nitrate before hurtling down the Pan-American Highway.

That night, the only place with any shelter from the desert winds was a large hill some way from the road. I struggled to push my stroller over the ragged earth to get as close as possible and set up camp on the hard bumpy ground. My water supply had run so low I had to hold back the desire to drink it in favour of cooking an evening meal. Despite the isolation, the setting sun filled me with joy as it turned the sky into a blaze of pink. The days required real grit and the conditions were genuinely harsh, but in a strange way this only added to my experience of the Atacama.

The road stretched south in long straight sections broken by shallow turns and at no point could you see anything but the tarmac disappearing into the flat horizon. Running here assumed a robotic quality in the unbroken sameness of it all. But as the day drew on, some hills appeared, and they housed a regional border control centre. This was not without benefit, as a small on-site restaurant produced a delicious three-course meal over which to research the road ahead.

One point in my map shone out. The small town of Quillagua is an oasis on the River Loa and, according to the Dirección Meteorológica de Chile and Guinness Book of Records, is the driest place on earth. It was a few kilometres further south and I decided to make my way there and look for somewhere to camp. The town itself reminded me of something out of The Walking Dead. Old wooden buildings seemed deserted as the wind hurled dust along the street between them. A donkey freely roamed the streets. This place's eerie feel was only broken by the sight of a family enjoying ice creams in the town square, and of old men sitting under trees, shaded from the sun. I asked around about a hostel or shop and got very little information. A conversation with a local lorry driver yielded slightly more, advising me that the checkpoint was the safest place to camp, which caused me to backtrack in search of the camping spot.

Like the rest of this part of the world, the ground was rock-like and without a drill, there was no way of getting my tent pegs into it. I searched everywhere and, in the end, with the permission of the restaurant owner, settled in the front garden of a rundown hut. It was the perfect size and, to the amusement of those passing by and a couple of stray dogs, I managed to squeeze my tent into the tight space, tying the guy ropes to fences. Back at the restaurant, when I gazed out of the window it was clear that the wind had picked up dramatically. The restaurant was located in a small valley and the tight turns and high sides created a funnel for the ferocious wind. Growing concerned, I raced back to find my tent jumping all over the place in the absence of anything to pin it to the ground. The only solution was for me to get in the tent to ensure it didn't fly away. With sleep not an option, I pulled out my laptop and watched a film as my tent cavorted around me.

With very little sleep, the start of the next day was a continuous struggle and lots of little things went wrong. But having been in this situation numerous times before, I didn't get too wound up and found it best to set about fixing each problem in turn. By the time I got back on the road I was in better spirits and some impressive petroglyphs and geoglyphs provided a perfect distraction. Not having showered in four days, I was also pleased to find a river in which I could wash and cool down.

As the day wore on, I started to find my rhythm again and the kilometres ticked by. Although people had told me there was nothing for hundreds of kilometres, I was pleased to find a few posadas along the road where I could replenish my water supplies. Never knowing where the next water would appear meant that I also tried to keep a decent amount in my stroller, even anything up to 10 to 12 litres. Certainly, I needed more than usual just to deal with these hot, dry conditions.

The exposed and baked Atacama is not the ideal place to pitch camp and this issue may on balance have been even more worrisome than remaining hydrated. After running 60km the next day, I stumbled upon a small restaurant but it appeared decidedly closed. This was private land, so I needed permission to camp, so searched around the surrounding buildings but everything was locked up. Not wanting to trespass, I was preparing to move on when two girls arrived in a lorry carrying tables, so I asked them if I could camp and they replied that the owner should be along shortly with the keys. We waited and waited and were about to abort when he finally arrived. After setting about unloading the tables into the hall, I was rewarded with a spot to camp and a couple of energy drinks.

On waking the next morning, I was thrilled that the restaurant was open and serving breakfast. As I was enjoying the typical fried eggs and crusty roll, blood dripped onto my plate in bright red globs and I realised that my nose had started to bleed. At only 1,000m, I couldn't attribute it to altitude so assumed it was due to the hot desert air. Shortly it was assuaged, and I was back on the road.

CHAPTER 31 - OVER THE ANDES

DISTANCE TO BUENOS AIRES: 2,212KM

Today was a momentous day for me as I was going to be leaving the Pan-American Highway that I had broadly followed since starting my run. Today the general direction of my adventure was going to veer to the east and towards the Andes Mountains, by far the largest of my obstacles.

The road towards the Andes came with new challenges. Gone were the long deserts flats and in their place was a slow upward incline directly towards the mountains in the distance, which to my dismay were not in fact the main Andes, just a range that lay between me and the town of Calama.

It was on this road that I had the most bizarre conference call of my life. I had just finished eating instant noodles on the side of the road, when I heard from Microsoft's PR agency. They wanted to confirm the details of the video shoot in Argentina. We went through all the high-level plans and then came to the specifics, most importantly where and when they needed me. The date of the filming was 28th November, which gave me 12 days to get there. The destination was Purmamarca, northern Argentina. My brain raced as they stressed how important it was that I was there on time as flights, accommodation and availability meant the dates could not be altered. I assured them I would do everything in my power to get there on time, though did outline just how big an ask it was. This meant covering over 600km in 12 days, crossing part of the Atacama Desert and traversing the Andes at altitudes over 4,800m. This would mean averaging 50km a day without any rests. I was expecting some sort of reaction but received nothing more than a straight answer of, 'yes, that's what we need you to do!' So, there it was, a challenge had been set and I had no alternative but to get as many miles into each day as possible.

While I ran, I tried to work out my plan for getting to Purmamarca on time. Also pressing was my very low water supply, certainly not enough to get me to Calama the next evening. There had been very few occasions on this adventure when I had misjudged my water strategy and there couldn't have been a worse place for it to happen than in the middle of the Atacama Desert. I started flagging down cars and lorries as they passed but the few that came by just kept on going. As the sun started to set, I felt rising panic and genuinely worried about having to stop until I found water. As these thoughts bounced around my head, one motorbike slowed down and to my relief, it carried a young Canadian traveller. He was also low on water, but his motorbike meant it would be easily obtainable the next day, so he generously shared his supplies and decided to camp with me.

The next day started with the final assault on the mountains and after a long struggle, I overtopped the 3,000m col and was greeted with my first clear glimpse of the actual Andes. I would cross them in a matter of days. Even from this distance they looked imposing but my focus couldn't be allowed to wander from the challenge at hand. Thankfully, an urban sanctuary of sorts awaited me that evening.

Calama is one of the largest cities in the region and is close to Chuquicamata, the world's largest open-pit copper mine. In this busy town, the stark divide between the rich and the poor was more than evident. I spoke with one of the locals who informed me that the mine was a magnet for people coming to find work. Infrastructure isn't there to support this influx and there are lots of schools but no colleges, so when children finish they have few options and as a result, alcoholism, prostitution and suicide are all very common. One of the big complaints is that the money made from mining doesn't stay in the region and therefore Calama itself doesn't benefit.

A more personal issue for me was finding somewhere to sleep. Having covered 600km over the last 11 days and with the Microsoft challenge front of mind, I was determined to get as much rest as possible. None of the hostels had availability or a room where I could keep my stroller. My mood dropped as I searched the streets and frustration swelled in my chest. In the end, I decided to once again throw money at the problem and pulled out the credit card for a room in a decent hotel. My room was in a colonial-style house and was worlds away from anything I had recently experienced. The sheets were white and the bathroom immaculate. With my wallet open anyway, I also treated myself to a proper meal in a steak restaurant close to the hotel.

On the way out of Calama, I stocked up on strawberries, cherries, apples, oranges and bananas from a huge fruit stall by the highway. I was going to need all the energy I could muster for the 100km I needed to cover in two days, pushing a running stroller weighing over 40kg. A little further on, while taking a break to enjoy the dramatic views, a campervan pulled over and out jumped a French traveller. He was one of those dramatically expressive people and was completely amazed at what I had taken on, and how far I had already got. Meeting people like this sparks the fire inside me and while only a short exchange, talking to him gave me the burst of energy I needed to keep moving forward.

I managed about 47km before being forced by the combination of exhaustion and the steep incline to resort to a march for the last 8km. Camping up on the pampa was an interesting challenge. Setting up a two-person tent in strong winds was more like flying a kite, and even harder with my body and mind sapped of all energy. My old enemy, frustration, was starting to build inside me as I continuously tried and failed to get my tent pegs in the ground. But in this bout, determination prevailed and with the sun dipping into the horizon I crawled into my tent to cook my evening meal. Inevitably the moment I finished the wind dropped to complete calm and I emerged from my tent to witness the most dramatic display of stars. There are many moments on an adventure when frustration can get the better of you, and I was no stranger to stamping my feet and shouting at the top of my voice. In moments like these, being solo is an advantage. Alone, you are free to express anything without judgement or the risk of affecting anyone else. Once the shouting is over, calm prevails and everything returns to normality. If I had been with someone else, then that anger could transfer to another person and engender prolonged aggravation. These are the kinds of interactions that can damage even long friendships. Sitting alone watching the stars, I felt blessed to be here—alone.

The splendour of the landscape I ran through the following day could rival anything on the moon. Giddy with excitement at the prospect of entering San Pedro de Atacama, seen as the gateway to the Andes, I powered on towards the town knowing there would be endless treats due to the vast number of tourists who base themselves there.

Even in the face of what lay ahead and the required daily kilometres, I still took a rest day in San Pedro de Atacama to prepare my body for the huge physical challenge ahead of me. Over the past 13 days, I had covered more than 700km at an average of 55km a day.

With a day off I had seven days to cover the last 420km, meaning an average 60km per day, with a vertical ascent of nearly 6,000m and running at altitudes reaching 4,830m. To add to that, my running stroller would be especially heavy. This stretch of road had very little potential to restock water or food, so to ensure success I had to push everything I would need for the crossing. The options were to take longer and push more weight or run faster and push less. With the pressing deadlines, I opted for the latter. My biggest concern was that there is no fresh water in the mountains as the water there is very high in salt. I had one day to come up with a solution.

As I walked around the town, I noticed that several tour operators ran tours to where I was planning to run. If I could persuade one of them to drive some water up into the Andes, that would dramatically reduce the weight of my stroller, without really causing any inconvenience at all to the driver. It took a while to find a tour that coincided with my run but I finally made an arrangement with Maxim Experience, who agreed to take two six-litre bottles of water to two different locations. The first drop-off would be at 4,600m and I would collect that on the second day. The next drop would be a further 50km along the road. Once again, the difference between success and failure in extreme conditions frequently comes down to problem solving and working with people to find solutions.

To add to the huge physical and logistical obstacles, some mental hurdles lay in my path. Having spent 13 days alone in the desert, readjusting to a town full of tourists, both rich and backpacking, rapidly sent me into a downward spiral. I felt myself retreating and getting irritated by people. In my self-imposed hermit state, it was hard to interact and I shied away from any form of conversation, preferring to hide in my hostel room. Perversely, this helped me concentrate on the preparation for what was to come because my solution to how I was feeling was to address the huge physical challenge head-on.

Taking on calories became a priority, so I made a real effort to eat as many burgers, pizzas and crepes stuffed with cheese as I could. There was just one other thing to organise before I started my assault on the Andes and that was to buy thicker socks. The temperature in the high mountains would be dramatically lower than what my camping kit was designed for. I was expecting the temperature to plummet to -10 degrees at night and I didn't want to be caught out.

After an early and hearty breakfast, I made my way through the quiet streets of San Pedro and out onto the road that would lead me over the Andes. If what I was experiencing wasn't surreal enough, it got more so when I heard someone shout my name. In yet another of the unlikely loops of my journey, I turned to see the group of backpackers that I had hung out with in Quito three months prior. We were all a little surprised that I had managed to run the same distance they had travelled in buses during the same period. We chatted and we discovered that if everything went to plan, we had a chance of all meeting again in Buenos Aires for my arrival at the Casa Rosada.

The Andes were on full display in front of me with a long road weaving its way up into mountains. From about 2,500m above sea level, ahead there was a further 2,300m of ascent. Mountains are deceptive, however, and while clearly impressive, they somehow seemed small. With nowhere else to go, I started running towards them and after a while, realised that they weren't changing in size and that perspective was playing tricks on me. In the distance, a bus was climbing upwards. I watched it for a long while and it just didn't seem to be making any progress, which was when it hit me just how physically demanding this day was going to be.

My aim was to run as much as possible, but with the stroller weighing in at nearly 50kg, running was hard to maintain as the gradient crept up. Day one's objective was just to get as much of the vertical climb done as possible. The fact that this was going to take all day didn't dishearten me at all and excitement bubbled through my veins as I took in epic views of volcanoes rising pyramidlike out of the barren ground.

While the route was devoid of dwellings, there were still people to meet, all of whom were foreign. The first was a French couple who stopped to check I was alive when they drove past and saw me attempting to sleep on the side of the road. Nothing seemed weird here, not even a couple of Germans who pulled over and introduced themselves as VW/Audi emission testers on a mission to calibrate and test the cars at altitude. As work trips go, this really had to be about the best imaginable. They said they would be returning the next day and would try to bring supplies. Finally, I startled a Scottish/French couple who had just cycled from Bolivia and were gobsmacked someone was actually attempting to run this route, let alone solo. We exchanged stories and even though already mid-adventure, I spent the afternoon daydreaming about potential future cycling expeditions of my own. It's hard to explain, but I was becoming aware that this adventure was coming to an end and what came next was something positive that I could focus my mind on.

As hard as I tried, I couldn't quite reach the top on the first day and decided to set up camp at about 4,260m after just 37km. As the sun set behind the mountains, the temperature sharply fell and setting up camp became an urgent priority. While I didn't suffer altitude sickness, I was experiencing slight headaches nonetheless from the change in pressure.

As soon as the sun warmed the morning air, I was back on the move. With the gradient far too steep for the last 500m of the ascent, I had no choice but to settle for walking. As I neared the summit, the Maxim Experience bus arrived as promised and agreed to drop my water at the top to save me the extra weight.

Not only that, but at about mid-morning the German auto-testers pulled up and, true to their word, brought the most amazing gifts. We sat by the side of the road and shared olive bread, Coca-Cola, water and blueberries. They had also brought additional water and other drinks that they were going to leave along the road for me to pick up as I went. Of all the things I imagined would happen on this crossing, this impromptu picnic was not on that list.

Clearly, my appearance here was somewhat anomalous, as a group on a sightseeing trip to the salt flats stopped to find out what I was doing. The guide in particular was excited to see someone running this route as he had never heard of anyone attempting to do such a thing. Apparently, the locals regard those attempting to cycle it as insane!

Again, the scenery was out of this world. Stone trees created by the winds, huge salt flats and red earth made it feel like the face of Mars. When the surroundings are this extraordinary, the running becomes strangely bearable, even at crazy inclines.

So desperate was I to record it all, that I was willing to constantly interrupt my rhythm to stop to take photos of everything around me. On one such occasion, I reached for my camera only to find it wasn't there. My reaction surprised me as I started shouting at the top of my voice, almost screaming because I had just crossed the most spectacular landscape of my life and had somehow misplaced my camera and all my photos. I crumpled to the floor, tears flowing for the first time on my expedition. The daily distances, altitude and the pressure I was putting on myself were clearly starting to take their toll. The options ran through my brain. Due to the distances and my supplies, adding extra mileage was not a realistic solution and one to avoid if at all possible. My last memory of the camera was 5km back, which could mean a huge 10km diversion. On every other occasion I had lost something on this trip, I had returned to find it and today was not going to be an exception. Time was now the most important factor. Turning my stroller around and running back the way I had come, the wind that had been my ally was now fighting me head-on. After the first kilometre, a big red pick-up pulled over to find out if everything was OK. Possibly looking slightly crazed, I explained my predicament and they offered to put my stroller in the back and drive me to where I thought I lost the camera. After 3km, there it was, my camera miraculously sitting unscathed in the middle of the road, presumably having fallen from my cart as I was running. My saviours offloaded my stroller and left me to continue on my way. To be honest, I was a little miffed they didn't offer to take me back to where they picked me up.

My German friends hunted me down again to show me photos of where my supplies were hidden. The sign would be a letter J they had written in rocks to the side of the road. This cheered me on, and I remain incredibly grateful for their immense generosity and ingenuity. In yet another life-affirming event, as the day was ending I met up with the Maxim Group a second time to collect my next water supply. Now I felt confident I had enough water to get me to the border, assuming nothing else went wrong.

Up so close to the sky, the open expanse of Andean landscape was dramatically punctuated with volcanoes and salt flats. Being so exposed to the elements at 4,600m made finding a spot to camp difficult. My task was therefore to find somewhere sheltered before the sun set and the temperature plunged to way below freezing. That night, I slept wearing every item of clothes I had, in a vain attempt to retain some sort of heat. But sleep I did, despite the cold, with limbs worn out from the demands of the day.

Like a mountain lizard, by the morning I craved the warmth of the sun on my tent so I could make the final assault to the highest peak of my adventure. A summit of 4,830m lay only kilometres away, but reaching it required me to leave the road and clamber over rocks, just so I would be higher than Mont Blanc (4,808m). Psychologically, I had naively assigned quite some significance to this moment. It seemed to represent the beginning of the end of the journey and in simplistic terms, it was arguably downhill all the way to Buenos Aires. While this helped to motivate me to get to the summit, I would later discover that the home stretch would not be plain sailing.

The scenery continued to astound me and with the wind firmly at my back, I set a good pace. Collecting my final water stash meant I now had a surplus, but because everyone had gone to so much trouble to provide it, I felt compelled to carry it. How could I snub such generous gestures from my helpers? On my way down I met more bemused tourists struggling to comprehend finding a runner in such a remote environment. In the vestiges of the afternoon, the wind had picked up enough to carry the grit from the soil, which stung my face. This felt like a slightly cruel last farewell from mountains I had so embraced.

Thinking it was another 70km to the border, signs indicating that it was behind me filled me with dread that perhaps I was meant to have had my passport stamped in San Pedro. With so much time on the road and in such a desolate area, my mind kept inventing different scenarios whereby I would have to return over the jagged 160km I had just covered. As the sky started to darken, I finally caught sight of a small border settlement in the distance. The only thing that kept me going was fantastical dreams of what luxuries might await me on arrival.

The border post consisted of a barrier over the road next to a hut, where a guard sheltered, playing games on his phone. He looked both surprised to see me and perplexed about how I got there. 'You can't cross this border on foot,' he said in a gruff, almost annoyed tone.

'Why not?' I answered.

He looked at me and grunted with inexplicable logic, 'This is a vehicle only border,' and told me to stay put while he found his superior.

Given that the border was quite established, my fears of being made to return started to subside. Whatever the outcome, I knew I would have somewhere to shelter that night. The guard returned with a colleague and took me to the main offices where we rushed through the exit and entry process. I asked the officer who processed me if there was anywhere I could stay and he pointed to a disused house where I could pitch my tent. As a parting gift, he gave me a small packet with a sachet of coffee and some biscuits.

The dilapidated building had evidently been used as a shelter for numerous other travellers. Tonight, I would be sharing this hovel with a Colombian traveller on a bike heading west, back the way I had just come. From the state of his bike, equipment and general fitness, I worried that he was going to struggle against the elements awaiting him on the mountain pass. We tried to share stories, but the language barrier was just too wide. From what I could understand, he drew and designed jewellery shaped as stars out of wire and hoped to sell them to fund his travels. When conversation ran out, he turned his attention to lighting a fire inside our hut and alarmed me by using fibreglass from a scrap heap to fuel the fire. The whole building filled with toxic fumes, and he was bemused when I rushed out of my tent and started to put out his fire. He had no comprehension of the potential damage he was doing to his lungs and in turn to mine. Eventually, the fumes died down but not before attracting the attention of a police officer. When he saw two men in a hovel, he jumped to the wrong conclusion. 'Are you gay?' he asked. To which, we explained our stories and then retired to our separate tents to get some sleep.

CHAPTER 32 - FILMING IN PURMAMARCA

DISTANCE TO BUENOS AIRES: 1,899KM

Crossing the border into Argentina was a pretty big deal. Here, in the last of my 14 countries, the biggest physical challenges, the Atacama and the Andes, were now behind me. Even though I was still at an altitude of 4,200m, had 1,900km to run and just 38 days to do it, I woke up in good spirits. What I didn't realise is that many more challenges were waiting that I hadn't yet anticipated. However, my immediate focus was on covering the 250km to get to Purmamarca and meet the team from Microsoft.

At a small store I bought everything I could find that would give me the energy for the four-day journey. No matter where you are in the world, there are always things you can buy that surprise and delight you and here was no different. My treat today was a large panettone, that I washed down with a cold Coca-Cola.

The road from the border was flatter compared with that of the last few days but the wind had moved direction and was now hitting me side-on, making running incredibly difficult. It was so strong that it had the force to blow one foot across the other and trip me as I was running. Finding a consistent rhythm was near impossible and it was a constant mental battle to keep forging forward.

Campervans were clearly a popular mode of transport for people travelling the Americas, as I had met several already on my adventure. This time the inhabitants were an amazing French family. They had been exploring South America for a few years and after a short trip back to France realised that they wanted to base their new life in South America. Unable to tolerate the high expectations, constant political issues and rise of terrorist acts back home, they'd decided that Ecuador would be their new home. Before departing, they helpfully told me about a small hut about 50km away that not only provided shelter, but also had Wi-Fi and water.

The day's drama struck when I pulled my water bottle out of its holder only to watch in horror as one of my Union Jacks was picked up by the wind, spontaneously unfurled and flew across the desert terrain, like a bird taking flight. Instinct took over and I set off like a hare, determined not to let it get away. When you have very little that reminds you of home, small things take on an unexpected significance. Every time I was about to catch my flag, a gust of wind would yank it away at the last minute. I ran and ran and after about 400m managed to get a foot on it —relief washed over me as I made my way back to the stroller with a victorious swagger. This little success filled me with something extra that got me back on my way.

My descent from the rocky mountainous terrain gradually transitioned to a greener, high-altitude pasture. Vicuñas grazed on the slopes, unperturbed by the curiosity of a gnarled runner and his stroller, while small huts nestled in the mountains. Although undeniably harsh, the life here clearly holds some appeal away from the pollution and noise of the city. Taking a dramatic corridor through a cleft in the rock, the world opened out before me onto an open plain of yellow sand glimmering in the evening light as far as the eye could see. Somewhere on this plain reportedly was a tourist office and the prospect of shelter, water and potentially Wi-Fi, although I remained dubious.

As the sun set and my watch approvingly clicked over the 60km mark, I spotted a small building in the distance and a lady, dressed in typical Andean clothing, walking towards the road. This was my chance to intercept her and ask whether this was the tourist office. She confirmed it was and as she had already locked up for the day, I asked if I could pitch my tent next to the building to shelter there from the evening winds. She smiled, leading me to the buildings and unlocking the door to a room that obviously served as the office and kitchen. It was warm and inviting inside and she told me I was welcome to sleep there. Not only this, but she disappeared into another room briefly, only to return with a proper mattress. To me at that moment, this represented unparalleled luxury, and I thanked whatever munificent deity might have pulled strings to help out. After showing me where everything was, the lady turned to leave. I couldn't let her go without enquiring about the Wi-Fi and she pointed to the password. That secured a night of contact with my family to reassure them I had safely managed the passage across the Andes and that everything was OK.

Checking in with people back home can sometimes be hard. The combination of feeling so detached from everything with being so physically and mentally drained, meant that all I wanted was to close down and recover before the next big run. It's painful and guilt-inducing to cut conversations short when you know the people at the other end just want to know that you are happy and safe. In these moments I found it hard to find the words to tell them I just needed to be alone and focus on recovery.

I had over 190km to get to Purmamarca and only three days to do it. This was enough to get me up early and on the road the next morning, determined to get to the next town, Susques, as early as possible to allow enough time to rest as much as possible before two even longer days. It was right in the middle of a huge plain that I was suddenly struck with an urgent need to evacuate my bowels. Of all the places for this to happen, I could not be more exposed. Running became a desperate search for any sort of shield from the cars that infrequently drove by. A small mound of earth a little further down the road became my target and I ran there madly before frantically scrambling over it, just in time. Recurring stomach issues were once again making running long stretches like today even more difficult as I was already having to deal with dehydration and a lack of efficient calories. There is a kind of glamour to the idea of adventuring, and ultra running has fairly lofty ambitions in terms of terrain and distance, but the reality is that the most basic, and at times embarrassing problems are the ones that will stand in your way.

Regrouping, the road took me through the desert and up into a rocky section of tight valleys before finally providing a stopping point in the town below. This was my second time in Argentina and I mistakenly looked forward to more European-style towns equipped with shops and established restaurants. When I arrived, I was sorely disappointed. Susques was quite primitive, and sheep roamed freely around its adobe brick-lined streets. Although the streets, buildings and local surroundings were pretty in a rustic way, I couldn't deny that I'd imagined Argentina was going to be easier than the last few countries. First, however, was a pressing need to withdraw Argentinean Pesos as my supplies were running low and using a credit card here was not always an option. As I searched, locals peered at me with curiosity out of windows or as they hurried past. To my endless relief, finally an ATM allowed me to withdraw enough cash to resupply.

The only hostel had definitely seen better times but luckily a bed was available. Before crashing onto a serviceable bed, I went in search of food and found only one shop was open and the selection there was scant. Trying to buy anything nutritious or that could be cooked was nigh on impossible and I came away with soft drinks, chocolate and biscuits. It was quite an insight into the workings of a town in such an impoverished terrain, and I wondered how the locals managed to secure provisions. A small restaurant across the street sold me 12 mini empanadas, not quite the feast I had hoped for. Looking around amazed me at the mismatch between this town and my preconceptions of Argentina.

The road out of Susques was spectacular, with long, sweeping bends winding down deep valleys where scrub plants covered the ground and shaggy llamas wearing pink ribbons grazed by the side of the road. It was on such a road that an approaching car swerved to the side of a tight canyon with sheer rock walls and three guys jumped out. To my surprise, it was the crew for the upcoming Microsoft shoot.

We'd only previously met via emails and calls and these guys looked so out of place in this harsh environment with their groomed appearance and latest technology. The producer was Sami, from

1000Heads, Alessandro was from Microsoft, while the director already had his camera in hand and was scoping out shots. They came bearing snacks and treats that I would normally regard as too expensive, but I was willing to make an exception in the circumstances. Because trying to convince me to accept a lift was out of the question, our rendezvous was short as a lot of distance had to be covered in a very short amount of time. They assured me that I would be happy with the luxuries that awaited in Purmamarca and then left me to it.

Alone once more, the road stretched for about 50km over a vast plain. Donkeys grazed either side and smallholdings were dotted around, presumably supporting fairly menial subsistence lifestyles. In another weird encounter given the location, a small pick-up truck converted into a camper pulled over and a couple jumped out, apparently happy to see me. After they jogged my memory, I remembered that I had met them in Colombia months before. We excitedly caught up on events, again marvelling at how two such different journeys could meet twice on such a vast continent.

The road slowly deteriorated the closer I got to the Salinas Grandes, a huge salt flat that covers over 200 square kilometres and attracts thousands of tourists. Running here was almost impossible as the surface was now a corrugated track, bordered by off-white salt crystals that had dried to form uneven hexagons and stretched in every direction. A small restaurant with tables and chairs made of salt, guarded by salt statues of llamas, looked like a perfect place to camp. But the gruff-looking man in blue overalls I asked retorted a harsh no before abruptly turning his back on me. This left me in a bit of a predicament as the sun was rapidly tucking itself behind the mountains and light was fading while the temperature plummeted. There was nothing for it but to part-walk, part-jog the next 10km, until I finally found a disused mine where my tent was partially shielded by spent heaps of earth.

The final push to Purmamarca was a day of two halves. First was a 27km climb back up to 4,200m before a marvellous 35km descent to 2,400m. Head-down, I ground out the kilometres as I climbed that morning, knowing that this would be my last hill for a few days. As I stood at the top of the mountain, I surveyed the road descending 2,000m into the valley below, switching back on itself time and time again. This was without a doubt the most spectacular road I had encountered and it reminded me of the Italian Alps. The best part was that not only was it downhill all the way, but at the end was a clean bed, a shower and a dinner funded by someone else's credit card. The price, however, was the intervening battle with a heavily loaded stroller without brakes for four long hours of descent.

At about 5.30pm I finally pushed my stroller across the cobbles, and on locating the rest of the team, I collapsed onto a chair. Over the last 21 days I had run 1,135km with just one rest day, averaging 57km per running day, crossing both the Atacama Desert and the Andes. However, if I had expected some kind of hero's welcome, that wasn't what I got. Sami looked at me and asked, 'Do you want the good news or the bad?'

'I think I need the good first,' I responded. He smiled, 'We have booked a delicious restaurant for dinner.'

'And the bad?'

'The alarm is set for 4am tomorrow morning.'

CHAPTER 33 - THE LAST BIG PUSH

DISTANCE TO BUENOS AIRES: 1,644KM

Filming with the Microsoft team was a surreal but welcome break from my usual routine, especially after a gruelling slog across the Atacama Desert and over the Andes and with very little rest. The hotel they had booked was certainly boutique; I was surrounded by people who spoke English as a first language, and all my meals were more gourmet and paid for. It was tricky trying to transition from being the solo adventurer completely in control of every aspect of my routine, to suddenly being at the command of a group of people I didn't know and had no inkling of how hard it had been to get there. After this, getting back into the groove was going to be hard and, with about 1,650km still to run and only 32 days remaining, this was not over yet.

Filming finished on the morning of 30th November and after farewells, I was dead set on reaching a campsite in the town of Yala, just over 50km away. My body felt rusty after the two days of standing around, but it was my mind that was struggling to readjust after such an intense dose of socialising and concentration. The next day I ran a mere 16km to San Salvador de Jujuy and called it a day. I needed to restock for the long road south, but also needed to reset my concentration and get into the right mindset.

Despite having two days off the expedition, working in Purmamarca wasn't really restful and small cracks were forming inside, both physically and mentally. My left knee was starting to ache, probably as a result of the long downward stretches where I had to hold my stroller back. My lower lip had also blistered so badly from the sun and wind that when I woke up in the morning my mouth was glued shut. It is no joke having to prise your lips apart in the morning before even being able to brush your teeth or drink water. This would then cause bleeding that made eating or drinking anything painful. My mind was also starting to buckle under the pressure of having to be in Buenos Aires for New Year's Eve. But it wasn't just that, I was starting to realise that this adventure was drawing to an end and I hadn't spent any time considering what came next. The future was opening up after months of knowing exactly where to go each day, and I have to admit that was very frightening. I needed some time to address all these imbalances and be in the best place possible to take on the mammoth challenge ahead.

A proper rest day did me a world of good and I felt confident enough to finish the job. I think it is worth pointing out that this next section of running was the hardest to write about and that attests to just how demanding it was. Raking back over the memories, I find huge gaps where my brain has clearly intervened to blank out the ordeal. My notes and videos are hard to watch, as they remind me just how gruelling it was. My strategy at the time was to attempt to trick myself into believing that this section was going to be easy, especially as all the big hurdles were behind me. The stroll into Buenos Aires would be casually triumphant. But what lay ahead would test me to the limit and draw on all the experience I had gained in the months and thousands of kilometres behind me.

Pushing my stroller out of Jujuy I had 1,575 kilometres to run and just 29 days to get there. This required an average of 55km every day without a day off, not even for Christmas Day, and to achieve this I would have to go to a place in my mind where nothing could distract or discourage me. I knew there would be challenges but this was my moment.

It was only a short distance south of Jujuy when the first issue arose. The primary route that connected the larger towns offered little opportunity to get away from the brutality of the highway. The roads were long and straight and massive lorries hurtled along them with little compassion for a solo runner. The noise, smell, fumes and heat in these conditions were beyond punishing. To make matters worse, there was very little hard shoulder to provide any sort of safety.

My first day was long and I managed to cover just shy of 70km before arriving in General Güemes and collapsing into bed in a hotel on the side of the road. The town itself was nothing to write home about but I was resigned to the fact that sightseeing was now firmly off the agenda and all my evening stops would encompass little more than refuelling and sleeping.

The next day revealed just how remote this part of Argentina was, away from the glamour of its lively cities. Either side of the monotonous road, unremarkable trees and fields filled the horizon while the small settlements became scarcer and scarcer. That night I pulled up under a bridge on the fetid banks of the Rio Pasaje O Juramento. As soon as I set about pitching camp, small black flies appeared out of nowhere and swarmed around me, crawling over any bare skin and invading my ears and eyes. Horrified, I used all the items of clothes in my possession to cover every patch of skin, not caring how ridiculous I looked. In my exhausted state after 56km, preparing food to replenish my energy was far from easy. My knee was playing up and with 27 days to go, I was starting to fear this pace would be unsustainable.

More disappointment awaited when venturing out of my tent to find the small black fiends were awaiting my arrival. Clearly, I was nothing short of delicious. Unable to tolerate the nuisance, I had to get running without any breakfast as standing still wasn't an option. If I thought that things would get easier, then I was to be sorely disappointed. Getting back onto the road produced a puncture and I can relay that it is very difficult indeed to repair an inner tube while flies launch relentless attacks on you. When I finally arrive at Metán after 48km, I am tired, itchy, ravenous and about to face more challenges. The first hotel had no vacancies and the next wouldn't accept credit cards. A third wouldn't allow my stroller in the room. At a street corner, my emotions were becoming my main enemy.

Frustration built up and the realisation that this was only day three of 29 was a brutal truth.

After finally finding a small hostel with space for me, I awoke the next morning in a more positive frame of mind. My mood only improved when I finally laid eyes on the first signpost for Buenos Aires. It announced that I had only 1,340km to go! Seeing my final destination there in writing, almost daring me to get there was just the mental lift I needed. For months, I had been running to a place that was too far away even to warrant a signpost and then all of a sudden its name was there in front of me. This new physical reality gave me something to count down towards. But there I went again. Silly me. While the end almost felt achievable, experience has taught me not to get complacent.

That afternoon laid another obstacle in front of me—a lack of cash. During my brief time in Argentina, it had become clear that cash machines rarely had available funds, which normally meant visiting every cash machine in the vicinity. The only problem was that those cash machines were getting further and further apart and my resources were getting low. On this particular day, I had arrived in the town of Rosario de la Frontera, home to what was apparently the last ATM for the next 240km. With just $20 in my pocket this was a crisis, as only a few places would accept a credit card. I duly tried each of the three ATMs in town but had no luck at any. Speaking to a cashier at the bank, she told me that they were expecting a delivery the next day but until then there was no way of getting money. This left me with two options. First would be finding a hotel, staying the night and praying that cash would arrive the next morning, or alternatively I could just trust in the journey and continue on with a mere $20 in reserve.

I weighed up the possibilities and realised that neither was ideal, so in the end it was my constant need to keep moving that prevailed. It was a strange feeling to cast out into the middle of a country you don't know with no money, knowing that the next ATM is four days' running away. Once at said cashpoint, there was no guarantee it would have any cash to give me.

To make matters worse, I was now joining the Ruta 34, or as many gloomily called it, the 'ruta de la muerte', due to the high level of fatalities each year. Its other main claim to fame is its role as the main drug trafficking route from Bolivia to Buenos Aires.

The next few days were soporific and it was all I could do to remain alert over 60km a day through huge expanses with nothing to really punctuate them. The odd small restaurant here and there sustained me with food and supplies when they accepted my credit card and, in the evenings, insect plagues thankfully far behind me, I camped on the side of the highway. My companions here were wild donkeys that I quite liked, black and red banded snakes sunbathing on the tarmac that I avoided, and clouds of white butterflies that I ran through, as if planted there by Walt Disney himself. As is their wont, ants invade everything and I had to brush them off my sandwiches before I contemplated eating them.

The heat was also rising and the humidity was getting unbearable. By midday, the mercury touched thirty-five degrees and I was desperate for anywhere that would provide shade or even better, sell a cold drink. With the increased heat came torrential rain, and that added another layer of complication. One morning the wind picked up and I could hear the patter of raindrops hitting the flysheet. Running through the rain is actually a good option, preferable to the scorching weather I know will follow. At this point it almost felt like my adventure was purposefully setting out to test my resilience, but I remained determined not to give in.

While the road was testing me to my limits, there were still short interactions with people that helped keep my spirits lifted, but I dare say that some dark corner of my soul started to enjoy the adversity. Shopkeepers gave me free food, cars stopped and gave me water and restaurant owners even ran after me with fresh empanadas or large chilled bottles of soda. While the road was testing me, its inhabitants were sustaining me and the draw of La Banda, the next big town was tantalising.

La Banda itself is nothing remarkable and perhaps it is a victim of urban sprawl, as it mainly comprises a patchwork of low-rise houses on the outskirts of its neighbour, Santiago del Estero. To me it was a supply town, a place to replenish my cash reserves, fuel up and get a proper night's sleep before battling on.

One strange distortion was that the distances between places had become difficult to comprehend, and the next morning I struggled to process the 500km gap from here to the next town of any size and while I may have struggled with the vast openness of the lofty plains, it was the flatness that really affected me. From La Banda to Buenos Aires is just over 1,000km and over that distance there is a mere 550m of elevation gain. For days and days, the horizon was just a haze in the distance, and with no natural landmarks to distract me, the road just faded into nothingness as all the while, the temperatures edged even higher.

The highway south of La Banda lost its hard shoulder altogether and became a tight two-lane road, seemingly without end. Lorries and pick-up trucks shot past, showing no sign of slowing down. With my back to the oncoming traffic, I was constantly startled by horns ordering me off the road. I had to form a strategy and every time I saw a lorry coming towards me, I made sure nothing was coming from behind. The near misses were adding up and stress built up tangibly in my muscles.

After 520km in just nine days, my body started to give way and I had intermittent diarrhoea to deal with on top of everything else. In the evenings, there was no choice but to wait until late before my tent was cool enough to sleep in, only adding to the sleep deprivation. I lay there listening to the cicadas and trying to anticipate what challenges tomorrow would present.

It was eight kilometres into the next morning before I realised that I had lost my pump. I could only assume that I left it where I had camped the night before. Over the last few days, a slow puncture had been an irritant at the periphery of my awareness, but I'd been too exhausted to actually fix it, preferring to keep it pumped up instead. It's so true that, when you are at your most depleted, is when you are most prone to these kinds of lapses. I knew I had no choice but to retrace my journey and find it. While running back the way I had come, of course my phone ran out of battery and as the temperature tipped over 40 degrees I felt a mix of desperation, rage and helplessness building up inside me. In this moment there was nothing I could do but keep calm, move forward, or backwards in this instance, and deal with each obstacle as it reared its frustrating head, but that didn't stop me screaming my anger into the void around me.

On 14th December, my 62km day ended at the small settlement of Herrera. On the outskirts was a small garage with a building that claimed to be a hotel, but more excitingly, a restaurant that advertised parrilla, or grill, in large letters on the roof. Trying to refrain from salivating, I was greeted by a bunch of local guys who worked in the garage. We got talking and somehow, I ended up joining them for a free meal. Argentinian steak tastes good at the best of times, but when you have run an insane distance in the oppressive heat and it's free, it's even more delicious. Tired and with a full belly, I ended the day there and opted for a room in the hotel. This room may have been basic, but when the torrential rain started hammering the roof I thanked the heavens that I wasn't outside.

The next morning, on my way south it immediately became apparent how bad the rain had been. Wading through flooded roads to the roadside restaurant I had set my heart on ended in disappointment when the staff were carrying buckets of water out and even ripping up the damaged floor. When I spoke to the locals, they told me that huge hailstones had caused major damage. Time to explore my second choice for lunch.

Over the next few days, I ran through many small and workmanlike towns, including Casares, Malbrán, Selva, Arrufo and Palacios before I finally arrived in Rafaela. I wish I could report that this 370km stretch was distinguished in any notable way, but I would be lying. It was mundane in its flatness and homogeneity. The main thing was to reach the town, and even the noticeable grinding of the ball bearings in my front wheel was not going to stand in my way.

There was a sense of relief when I arrived in Rafaela because my 13 days of running along the hellish Ruta 34 were over. Nearly 800km put to bed, just like that. This section tested my mettle in so many ways, as nearly every niggle that I had experienced over the rest of this adventure also happened on this short stretch, except the distances I was covering were far greater. Every day, I would get up and run along a long, straight road with lorries bearing down on me under the oppressive heat. I had no idea if the road south of here and the remaining 570km would be any different, but I felt a sense of hope return. The Ruta 34 had left me feeling numb and nearly broke me, but now I had come to the junction where I could leave it behind. The feeling wasn't tiredness or defeat, instead I felt triumphant and excited for the next and final push to Buenos Aires.

With just 570km to go and just 12 days to do it, I started flirting with the idea of a day off. My mind and body were tired after running 1,000km over the last 17 days but reason got the better of me. What if something went wrong? I couldn't let myself fall short this close to the finish line. In the end, the compromise was to let myself have a later start and take advantage of the Wi-Fi to address some real-life admin, such as a researching flights home, transferring money and even paying my UK car tax. By now, there really were two Jamie Ramsays, each vying to win the debate, one begging to relax, and one strictly enforcing the deadline.

It didn't take long for the joy of being away from Ruta 34 to be cruelly dashed. After enjoying the quieter and prettier roads, my stroller suddenly came to an abrupt halt as the ball bearings on the front wheel finally seized up. I was stranded, miles from anywhere without the necessary tools to fix the problem. All that could be done was to push the handlebars down to raise the front wheel and continue running with the stroller in wheelie mode, which was far from ideal. Then, with the innate sense that bullies possess, a chap I had never met sent me a message at this moment of weakness that simply stated I was 'a fucking idiot'. Stranded in the middle of Argentina with a stroller that wouldn't move, I have to admit, I did feel a little like one!

In the small, quiet town of Neuvo Torino I stopped for supplies and got talking to the shop owner, who gave me a bottle of ice-cold water. He introduced me to one of his customers, a chap called Adrian who, on hearing about my wheel, immediately invited me to his house where he introduced his family and promptly started to take my wheel apart. Soon, it was spinning freely once again.

As the sun set on yet another day, I met some cyclists who invited me to stay in their village, Pilar. When I got there the pretty main square was abuzz with locals keen to enjoy the cooler evening climate. The hospitality was exemplified by the jolly hair-netted owner of a fast-food van parked nearby, who gave me a burger for free. I took so many selfies with random people that night and remember that humbling feeling of being the traveller who has arrived in a quiet little village where everyone is happy and accommodating. It topped off a day that re-affirmed just how much I enjoyed being out there on the road, experiencing so many small reminders of human connectedness and warmth.

The days leading up to the finish seemed to fly by and the surging anticipation made me feel like the end was so close I could taste it. But I also knew that back home everyone was with their families and preparing to celebrate Christmas, always a hard time to be alone and constantly on the move. Once again, I think subconsciously I made my days even busier to distract my mind from what I would be missing.

News had reached me of a hotel in the town of Barrancas, and at that point a bed was exactly what I needed after a few days of incredibly hot running and a few nights struggling to get any decent sleep. By now, my tent was acting more as a sweat room than a place of rest and recovery. But my hopes quickly turned to sadness as there was no room at the inn. As I sat in a local bar eating cheese and ham sandwiches, crisps washed down with Coca-Cola, I spotted the local fire station and wondered if they would be accommodating to a weary traveller, having been tipped off many months previously that it sometimes worked. Opening the big metal sliding doors, I found a group of volunteers in smart blue uniforms. The members of the group varied in age but all had that vibrant sense of being good. When I asked if I could camp, they all got very excited and went out of their way to help. They showed me a perfect spot for my tent in the garden, before we all sat down and they bombarded me with questions about my journey. This little moment of telling tales from the road, especially my Argentinian experiences, fostered incredible catharsis. It allowed me to sort through my experiences, and from other people's reactions actually take stock of my achievement and absorb it in a way that I couldn't while slogging it out on the highway. They were all very pleased when I declared that the most beautiful girl I had met on the whole journey had come from the village some 15 kilometres north.

On Christmas Eve I arrived in the beautiful city of Rosario. After nearly 1,600 kilometres of running in Argentina, I was finally standing in a town that matched my mental image of this nation. Grand buildings on tree-lined avenues, green parks, museums, art galleries and the national flag flying everywhere. This town had more than a tinge of Europe about it, but even so, I didn't feel like I belonged here. The noise was never-ending, with people and activity everywhere. My senses were overloaded with so much commotion and immediately I felt a need to escape. So many people were telling me to take a day off for Christmas, but my body wanted nothing more than to continue moving and to get away from the hustle and bustle.

So, on Christmas morning, I packed up my stroller and made my way through the hugely impressive Distrito Centro and the Parque Independencia and back to the highway where I seemed to belong. My quiet highway soon morphed into a full dual carriageway, and I feared that the police might have something to say about a solo runner. As it was Christmas day, I forged on, trusting that it was unlikely that anyone would pull me over.

My Christmas dinner was spent on an island but sadly not a tropical one. This was the AXION Energy petrol station between the two carriageways. My feast that day was microwaved gnocchi and chocolate cake before venturing back onto the highway. That afternoon I felt like Joseph in some nativity play, endlessly searching for somewhere to sleep. As Buenos Aires was getting closer, available wild camping pitches were becoming fewer and further apart. In the end I pushed on for over 70km until I saw a sign for San Nicolas and took it as an omen. Unsurprisingly, not much was open until finally refuge was possible in a small hotel near the centre of town. I drifted off to sleep that night with the comforting thought of having just 250 kilometres to cover and six days to do it. Closure of this adventure was within my grasp, and not only did I realise I was going to make it, but I might even get there a whole day early.

I am very grateful that I managed to squeeze in two more days of camping before the sprawl of Buenos Aires took over. They may not have been the prettiest places to pitch a tent, but it allowed me two nights of reflection on the journey that was fast coming to an end. It had been 500 days since I had first started running in Vancouver and here I was, about to run into Buenos Aires, having completed something that many thought impossible and even more couldn't comprehend. After all those nights in this tent, I took one more photo for good measure on one of those last nights. It captured a view that is so special to me; my stroller parked in the entrance of my tent with my Union Jack hanging down.

When recalling the thoughts running through my mind in my tent that night, I smile at how different it was from what people might imagine. Rather than revelling in the rich and varied experiences along the way, for some reason I spent a lot of time thinking of all the missed opportunities instead. I chastised myself for setting such an ambitious target of getting to Buenos Aires for New Year's Eve and dreamt about all the things I could have done if I had slowed down. I worried about arriving a day early and whether I would have to delay it if anything happened to my girlfriend's flight. It even troubled me that I had become so accustomed to eating over 5,000 calories a day because I had to, and now this adventure was over I would have to eat normally again.

More stark though, was the notion that I had absolutely no idea what I was going to do when this all came to an abrupt end in just a couple days. I had no plan, no money and no direction. I was going to be left with very little but a very good story to tell. The aim of this adventure wasn't to make a documentary or break any records; I had done it for me.

It was during these quiet moments alone that I made the decision to prolong my break from alcohol. I had heard of adventurers suffering from post-adventure depression that led to long spells in the pub regaling people with stories. I knew my personality, I knew my weaknesses and I knew that to preserve my sanity and to hold onto the changes I had made, I would have to stay on the straight and narrow, slowly reintegrating myself back into life on the other side of the adventure.

On 29th December 2015, I arrived at the Hotel Munro in the north of Buenos Aires. I knew I could have reached the Casa Rosada that day and would have been two days ahead of schedule, but it would have meant nothing if my girlfriend wasn't there. There was no quest for glory and to slow down a little before the end was probably the right thing to do. That evening I held an emotional ceremony to get rid of all the items I had carried for thousands of kilometres but were now redundant and in many cases no longer worked anyway. I threw away my water filter bags that were taped up with duct tape, the sleep mat that had split and the pee bottle that had faithfully served me throughout.

No last day would be complete without a drama. My passport had been vital for checking into hotels all across Argentina, but as I did my final pack it was nowhere to be seen. I searched everything, throwing my possessions everywhere in a frantic panic. After alerting hotel security, I pleaded with the front desk to help search the hotel, confident it must be somewhere. At least in that regard, I was correct, as a short while later I had to sheepishly return to tell everyone that I had found my passport and it was actually exactly where it was meant to be, just hidden behind some other papers.

With until 3pm to run the 20km to the Casa Rosada, I took my time and even visited some tourist spots along the way, including the Chacarita Cemetery, a sprawling necropolis renowned for its elaborate mausoleums and final resting place of many notable figures in Argentine culture and history. In fact, so absorbed was I in exploring the city that I lost track of time and my phone battery died amid the plethora of photos I took that day. To catch up, the last few kilometres were spent dashing from street to street, dodging traffic jams and having to navigate around one-way systems. This panic added to the adrenalin that was already coursing through my veins as I made my final turn on to the Plaza de Mayo and the Casa Rosada, where Eva Peron gave her speech in 1951. While running, my eyes roved the park for my girlfriend. When I caught a glimpse of her, I whistled loudly to catch her attention and then made the final strides of my adventure. With a closing fist pump, that was it, in the briefest of moments it was all over!

My girlfriend was not alone, she was accompanied by the three backpackers that I had met in Quito and then again in San Pedro de Atacama, all waving homemade signs with slogans that variously read, 'You run better than the government' and, 'I want your babies'. A couple of reporters loitered and a photographer snapped a few photos.

I am not sure what I had expected to feel in that moment and had really not spent much time thinking about it. There isn't really a way to prepare for the culmination of an event like this. But once again there was no wave of achievement, no sense of completing something heroic or anything like that. Just half an hour spent chatting in the sun, then I suggested we head to a bar to grab a Coca-Cola. The adventure was over. Just like that, after months of running, thousands of kilometres of pushing myself to my mental and physical limits, in a single second it was over and there was peace.

I will admit that there was a sense of pride in arriving in Buenos Aires a day ahead of schedule. In the end I had spent 177 days in South America and was forced to take 27 unplanned days off. This meant I ran 7,666 kilometres in 150 days, averaging just over 51km per running day. This little bit of maths helped to justify my sense of achievement. It's not winning, it's not setting records or achieving firsts. It's just about being out there pushing your mind, body and soul to a place you previously didn't think possible. It's the realisation that you can always do more than you believe is possible. That is why this adventure began in the first place because when I quit my job in April 2014, I wasn't proud of what I was achieving, I didn't have any sense of accomplishment. This adventure gave that to me.

Printed in Great Britain
by Amazon

60924486R00201